D0975301

NATURAL GAS
THE BEST ENERGY CHOICE

NATURAL GAS
THE BEST ENERGY CHOICE

Ernest J. Oppenheimer, Ph.D.

PEN & PODIUM, INC.

Natural Gas, The Best Energy Choice
By Ernest J. Oppenheimer, Ph.D.

Publisher: Pen & Podium, Inc.
40 Central Park South
New York, N.Y. 10019
(212) 759-8454

Printed in the United States of America

ISBN-0-9603982-7-9

Private Printing: Paperback Edition ISBN-0-9603982-8-7

Library of Congress Catalog Card No. 89-63205

Contents

Forewords

By Dr. Stephen D. Ban, President and Chief Executive Officer, Gas Research Institute

This informative book presents objective criteria, such as value per dollar of usable energy, environmental cleanliness, security of supply from domestic sources, and versatility to demonstrate that natural gas is a highly competitive option in all energy applications.

As the author points out in Part I of the book, we are fortunate in having plentiful natural gas resources in the U.S. If properly employed, these resources will take care of our needs for a long time to come. The Gas Research Institute is sponsoring research on advanced technologies to achieve improved ways to produce and utilize these resources, including methods to achieve cost-effective recovery of nonconventional forms of methane gas.

The uses of natural gas are closely linked to technological developments. In many areas, gas is on the leading edge of technology. The book gives recognition to this reality by devoting considerable attention to new developments in efficient gas heating, cooling, cogeneration, combined cycle electricity production, and vehicles fueled with compressed natural gas. Much of the research to bring these technologies to practical application in the competitive energy market has been sponsored by the Gas Research Institute on behalf of the natural gas industry and gas consumers.

The book does a highly effective job of communications on a technically complicated subject. The average reader should have no problem understanding the text, which provides a comprehensive view of a topic that is vital to the energy and economic future of our nation.

By Paul S. Hughes, Director of Corporate Relations, World Resources Institute

Natural gas can virtually eliminate pollutants such as sulfur dioxide, carbon monoxide, reactive hydrocarbons, and particulates. It can substantially reduce nitrogen oxides and carbon dioxide. Dr. Oppenheimer's book explains the environmental advantages of natural gas, its availability, and its capability for meeting our national energy needs in the future.

Industry experts expect conventional natural gas resources in the U.S. to last another 40-50 years at current rates of consumption. Even this projection is likely to be conservative, given that price and technological developments should enable the industry to extract natural gas from vast non-conventional sources. Natural gas is also a renewable resource. In nature, it is produced by anaerobic microorganisms, which ingest biomass and give off methane within 30 days. This process can be utilized to manufacture methane from manure, sewage, garbage, plant residues, and energy crops. Millions of peasants in India, China, and other developing countries are already making use of this ecologically beneficial technology.

The natural gas industry should experience significant growth in the use of combined cycle electricity generating equipment, cogeneration, cofiring with coal, and compressed natural gas (CNG) in buses and fleet vehicles. This growth is apt to come first in the hundred or more urban areas in the U.S. that are not meeting current ambient air standards required by the Environmental Protection Agency.

All the elements are in place for a surge in natural gas usage. The new environmental awareness by leaders of the natural gas industry will facilitate this process. They are now taking the initiative to advertise the environmental advantages of natural gas and to present the case for this fuel to state utility regulators in connection with generating electricity.

For the lifetime of most readers of this book, natural gas will be the medium which enables us to make the transition from an essentially nonsustainable energy-intensive society relying primarily on non-renewable energy sources to one which makes optimum use of renewable ones, such as solar thermal, photovoltaic electricity, micro-hydroelectric, geothermal, and wind sources.

The natural gas industry deserves greater attention and support by environmentalists and society as a whole. Dr. Oppenheimer's book makes a convincing case for this new perspective, which should result in closer cooperation between environmentalists and the gas industry.

By David J. Bardin, Partner, Arent, Fox, Kintner, Plotkin & Kahn, Washington, D.C. Mr. Bardin was formerly Administrator, Economic Regulatory Administration, U.S. Department of Energy, 1977-1979; and Commissioner, New Jersey Department of Environmental Protection, 1974-1977.

Debates about relative future roles of natural gas and other energy sources engage businessmen and investors, economists and technologists, analysts and editorialists, scientists and geopoliticians. Dr. Ernest J. Oppenheimer has examined the issues from the vantage of a social scientist, with experience in the ways of investment banking. Focusing largely on the United States, he marshals data and analyses from such recognized sources as the American Gas Association, the Gas Research Institute, and the U.S. Department of Energy to help him outline economic, environmental, technological, and strategic postures and prospects of natural gas as an energy source.

Abundance of natural gas resources is succinctly expounded, together with the dynamic processes by which undiscovered and unconventional resources become part of the immediate inventory on which the economy relies. Similarly, end use technologies (such as the increasing efficiency of gas generation of electricity, and the development of gas engines for vehicles which exploit the 130 octane rating of methane), create opportunities for increased natural gas usage.

This book offers a useful and timely review. Natural gas, and the million-mile underground network which delivers it, are generally invisible and may therefore go unnoticed. One hopes that Dr. Oppenheimer's handy exposition will generate controversy and debate about the virtues of natural gas for policy decisions as well as for public understanding and acceptance.

Preface

I have beeen actively involved with research and writing about natural gas since 1979. My first book on this topic, *Natural Gas: The New Energy Leader,* was published in 1981. A revised edition came out in 1983. Since then, I have published a number of articles on natural gas. I have worked on *Natural Gas, the Best Energy Choice* since 1988.

These studies have led to the conclusion that the increased use of natural gas can help achieve the following objectives:

(1) Avoid excessive dependence on imported oil.

(2) Improve the environment, including reductions in smog, acid rain, and the greenhouse effect.

(3) Strengthen the economies of the domestic petroleum producing states.

(4) Save consumers billions of dollars on annual energy bills.

Any one of these benefits would justify greater interest in getting acquainted with this fuel. The fact that natural gas can accomplish all of these positive consequences simultaneously makes a powerful case that should be brought to the attention of all those concerned with energy and related issues. This book is dedicated to the task of giving readers the opportunity to become well informed on this important topic.

Sources

Many sources have been used in the preparation of this book. My files of research materials on natural gas are filled with studies by the American Gas Association and the Gas Research Institute. I have also made use of U.S. government reports, particularly publications of the Department of Energy and the Commerce Department. United Nations studies were drawn upon for international data.

The American Gas Association is the main trade organization of the natural gas industry. It has a sizeable staff of professional researchers preparing studies on gas supplies, technologies, markets, and other topics. These studies are designed to inform members on current developments and trends affecting their business. Many of these studies were utilized in the preparation of this book.

The Gas Research Institute, which is financed by gas consumers, sponsors research on gas resources, technologies, and equipment. Its staff consists largely of scientists and engineers who supervise research projects. They also prepare reports on these projects. Several of these studies provided information for this book.

I would like to take this opportunity to thank a number of individuals for their helpfulness. Mr. George H. Lawrence, President of the American Gas Association, encouraged my independent study of natural gas and has been very cooperative in making available to me the research materials prepared by his staff. I have known Mr. Lawrence for more than ten years. I hold him in high regard for his comprehensive knowledge of natural gas and his realistic assessment of the gas industry's problems and opportunities.

I contacted Dr. Steven D. Ban, President of the Gas Research Institute, in connection with the research for this book. He was most cooperative in giving me the information I requested. I particularly appreciated his willingness to explain highly technical matters in language I could readily understand and pass on to my readers.

Many professional staff members of the American Gas Association gave me valuable assistance in connection with research for the book. The following individuals played a particularly important role:

Mr. Nelson E. Hay, Chief Economist and Director, Policy Analysis, coordinated my contacts with the A.G.A., sent me relevant studies, and was generally most helpful.

Mr. Robert B. Kalisch, Director, Gas Supply and Statistics, did an outstanding job providing me with up-to-date information for preparing Part I of the book.

Mr. Richard Itteilag, Director, Gas Demand Analysis, supplied much information and documentation for several chapters in Part II of the book.

I also benefited from the studies and comments of Mr. Bruce Henning, Manager, Economic Analysis; Mr. Leon L. Tucker, Director, Energy Modeling Services; Mr. Paul McArdle, Policy Analyst; and Mr. Paul Wilkinson, Associate Director, Policy Analysis.

The book's practical value was enhanced by the input of the following individuals from the Marketing Department of the American Gas Association: Mr. Mark S. Menzer, Vice President, Marketing Services; Mr. Philip J. Mahla, Director, Residential Marketing and Market Research; Mr. Mark Wilson, Manager, Gas Cooling and Cogeneration; and Mr. Allen Weidman, Director, Commercial Marketing.

Another valuable resource was Mr. Steven J. Dorner, Director, Industry Information Services, who is in charge of the A.G.A. library. Mr. Dorner's expertise on books, and his understanding of the needs of libraries, has influenced my thinking to make the book more useful for that important audience.

For the chapter on natural gas vehicles, I am grateful for information made available by Dr. Jeffrey Seisler, Executive Director, The Natural Gas Vehicle Coalition, and by Mr. Robert J. Fani, Manager, Natural Gas Vehicle Marketing, the Brooklyn Union Gas Company.

Mr. Joseph S. Collier, Vice President, Marketing, the East Ohio Gas Company, sent me useful information on gas cooling and cofiring. Mr. Richard J. Trieste, Assistant Sales Manager, Residential Marketing, the Brooklyn Union Gas Company, helped enlighten me on the advantages of natural gas over oil for space heating and hot water production.

While I value these contributions by others, the responsibility for putting this book together, and for its contents, rests with me. I found the task challenging. I hope the readers will get as much intellectual stimulation reading the book as I had writing it.

Summary

Chapter 1. The outlook for natural gas is favorable. Adequate supplies of this fuel are available from domestic sources, supplemented by imports from Canada. The environmental advantages of natural gas are increasingly recognized by the general public and by government officials. The 1,150,000 miles of gas pipelines and mains provide reliable service to 48 million residential, 4 million commercial, and 200,000 industrial meters, serving an estimated 172 million customers. The development of highly efficient gas appliances and equipment, and new technologies for gas cooling, cogeneration, and combined cycle electricity production, have broadened gas markets. The nation will benefit from the increased use of natural gas, which will strengthen energy security and help create a cleaner environment.

Chapter 2. Methane, the principal ingredient of natural gas, originates from the following sources: (1) Organic matter in sediments, whose decompostition is being promoted by heat; (2) The action of anaerobic microorganisms that convert organic materials into methane; (3) The transformation of oil and other heavy hydrocarbons into methane at high temperature, usually in deep locations; and (4) Coal, which releases methane as it matures. Some scientists believe that methane may have been present at the time the earth was first formed. This hypothesis postulates that methane may in part come from nonbiological (abiogenic) sources. The validity of this theory remains to be proven.

Chapter 3. The Potential Gas Committee estimates conventional gas resources in the U.S.A. at 983 trillion cubic feet (Tcf), enough to last fifty years at current rates of consumption. In addition, gas in tight formations is estimated at 2,800 Tcf, about three times as large as the conventional resources. The gasification of coal, peat, and oil shale could generate enough methane to last for a thousand years. Methane can also be produced through the action of anaerobic microorganisms from landfills, manure, sewage, and energy crops. As long as the sun shines and plants grow, human beings can obtain all the methane they need for the rest of time.

Chapter 4. Conventional gas resources include 187 Tcf proved reserves, 587 Tcf of potential resources in the continental U.S.A., 90 Tcf from coal seams, and 119 Tcf from Alaska. The largest amounts of natural gas are produced in Texas, Oklahoma, Louisiana, and from offshore locations owned by the federal government. Gas in locations below 15,000 feet and offshore under thousands of feet of water have promising prospects for adding to conventional gas supplies. Improved geological procedures and technological developments are likely to lead to increased conventional gas production at reasonable cost. They will also broaden the types of gas that can be placed in the conventional category, including methane resources currently classified as nonconventional.

Chapter 5. Large quantities of natural gas are trapped in tight sands, Devonian shale, coal seams, and geopressured brine. The more accessible gas resources from tight formations are already being utilized; they account for 1.4 Tcf of annual production. Additional recovery from tight formations is a function of technology and price. It is likely that these large resources will be increasingly put into production through advanced scientific know-how.

Chapter 6. Vast quantities of methane can be obtained from gasifying coal, peat, and oil shale. Gasification of solid fossil fuels is far more benign to the environment than burning them. In fact, many of the substances which cause pollution during combustion are recovered as valuable byproducts during gasification. Several processes for gasifying coal and peat have been developed. A large coal gasification facility is in operation in North Dakota. Greater utilization of gasification is likely in the future, when conventional natural gas resources are less readily available and become more costly.

Chapter 7. The conversion of biomass into methane by anaerobic microorganisms may hold the key to permanent energy independence. This process can be observed in landfills, where methane is released from organic debris. The same procedure can be applied to manure, sewage, and energy plant crops grown on land or in the ocean. These materials can be placed into biodigesters, large tanks that are closed off from air. Within a few days, methane begins to form. It has been estimated that 55,000 square miles of ocean energy farms could supply 20 Tcf (trillion cubic feet) of methane gas annually for the rest of time.

Chapter 8. U.S. consumers are served by more than 1,150,000 miles of gas pipelines and mains, backed up by storage facilities that can hold 7.8 Tcf of gas. This distribution network is adequate to take care of foreseeable needs in most areas, including considerable room for expansion. Additional pipelines are being constructed to take care of regions experiencing rapid growth, such as the Northeast. Natural gas from Alaska will probably be brought to the lower United States via Canada in the early part of the twenty-first century.

Chapter 9. Canada has very large natural gas resources, which can take care of her own requirements as well as allowing for substantial exports to the U.S.A. In 1988, Canada exported about 1.3 Tcf of gas to the U.S. These exports are expected to grow over the next several years and may account for as much as ten percent of total U.S. requirements.

Chapter 10. The long-term supply outlook for natural gas is positive. Production in the continental U.S. is expected to take care of most requirements well into the twenty-first century. Alaskan gas production is likely to become a significant source after the year 2000. Imports from Canada will take care of most of the balance.

Chapter 11. Demand for natural gas is expected to grow from 18.5 quadrillion Btu in 1988 to 23.1 Quads in 2010. Most of the growth will be in advanced technologies for generating electricity (cogeneration and combined cycle), gas cooling, and compressed natural gas for vehicles.

Chapter 12. Space heating accounts for the largest use of gas in residential and commercial markets. Gas furnaces and boilers achieving efficiency of 90 percent or more are now widely available. During the past sixteen years, conservation and improved efficiency have kept gas consumption for space heating fairly level, even though the number of customers has increased substantially. This trend is expected to continue. Heating with gas has significant advantages over competing energy sources. Gas is more convenient and cleaner than oil; it is much less costly than electricity.

Chapter 13. Gas is the most economical fuel for hot water production. It has been estimated that residential customers can save $200-300 a year by replacing old oil or electric water heaters with gas units in many parts of the country. Such savings can be achieved with a gas water heater costing as little as $500 installed. The annual return on investment can be as high as 40-60 percent.

Chapter 14. The gas industry is making a major effort to expand its share of the cooling market. A number of efficient new gas cooling products for the commercial market have been introduced recently or will be offered in the near future. Gas heat pumps for the residential market are also being developed. Gas cooling would bring about a more balanced year-round demand for natural gas, which is currently lopsided because of heavy use for heating during winter months.

Chapter 15. Cogeneration is a procedure for generating electricity that also utilizes the waste heat accompanying the process. Because heat energy cannot be efficiently transported over long distances, the cogeneration system must be located near the site of consumption. Packaged cogeneration units are available in a wide range of capacities. Cogeneration is well suited for many hospitals, schools, hotels, office buildings, and apartment houses. Natural gas is the preferred fuel for cogeneration. In 1987, about 800 billion cubic feet of gas were used to generate 17.3 gigawatts of cogeneration capacity. An additional 13.5 gigawatts of cogeneration are under construction or planned for the next several years. Cogeneration is one of the most rapidly growing markets for natural gas.

Chapter 16. Electric utilities face the following challenges: (1) They must meet increased demand for electricity; (2) They must reduce environmental pollution; (3) They must keep capital and operating costs under control. Natural gas is the choice fuel for helping electric utilities cope with these challenges. Relatively small quantities of gas cofired with coal can greatly reduce air pollution. Gas-fired equipment can be used to prolong the life of existing electricity generating facilities. Combined cycle technology, which makes use of waste heat to generate additional electricity, is an efficient and cost-effective procedure for constructing new electric capacity. This technology has the advantage of relatively low capital and operating costs. It allows for flexibility through modular construction and it can be ordered and installed in a relatively short period of time. Over the next several years, combined cycle technology will be a major growth market for natural gas.

Chapter 17. Commercial markets for natural gas have been growing in recent years and are likely to make further advances in the future. Gas is the preferred fuel for commercial space heating, hot water production, cooking, and drying. New technologies, including gas cooling and cogeneration, will find increased application in commercial markets. In 1988, more than four million commercial customers used 2.7 quadrillion Btu of gas.

Chapter 18. Between 1973 and 1986, U.S. industry *increased* output by 33 percent while *reducing* energy consumption by 23 percent. Energy efficiency has become a major consideration in virtually all manufacturing processes. In this period, the use of industrial natural gas declined by 3.6 Tcf. However, since 1987 industrial gas consumption has risen. New applications for gas utilize its premium qualities, including its cleanliness and its compatibility with highly efficient operations. Industry is making increased use of gas-fired cogeneration.

Chapter 19. Compressed natural gas (CNG) is an excellent vehicular fuel. It minimizes air pollution. In comparison with gasoline CNG eliminates more than 80 percent of the carbon monoxide, one-third of the nitrous oxide, and most of the reactive hydrocarbons. Methane has an octane rating of 130, much higher than even the best quality gasoline. Vehicles using CNG have a better safety record than those fueled with gasoline. Studies indicate that in case of collisions, the injury rate among passengers in CNG vehicles is much lower than in gasoline vehicles. This phenomenon may be largely due to the fact that CNG is a safer fuel than gasoline. CNG is stored in a cylinder that survives intact most collisions. Moreover, even if it were ruptured, the gas would harmlessly escape into the air. In contrast, the gasoline fuel tank is easily broken during a collision and the gasoline puddles to the ground and often catches fire, which increases the risk of injuries.

Chapter 20. Natural gas is the cleanest of all fossil fuels. Its production involves minimum disturbance of the surroundings. Cleaning of the gas *before* it is put into pipelines removes most pollutants. Natural gas can play an important role in reducing ground level ozone (smog), acid rain, and the greenhouse effect (global warming). These environmental advantages of natural gas are being increasingly recognized by the general public and by governments.

Chapter 21. Natural gas can strengthen energy security by reducing excessive dependence on oil imports. In 1987, oil imports accounted for about forty percent of total consumption. Because of declining domestic oil production, it is likely that the share of oil imports will increase to more than 50 percent by the mid-1990's and may exceed 60 percent after the year 2000. Natural gas can help counteract this trend by replacing 1.3 million barrels of oil per day in stationary applications. Increased use of gas for vehicles would also help reduce oil imports.

Chapter 22. For all practical purposes, natural gas prices are now determined by free market forces of demand and supply. It is believed that this approach will facilitate long-term price stability, with a modest uptrend as demand increases. Such a pattern would help broaden markets for gas while increasing supply.

Chapter 23. Gas and electricity have a complex interrelationship. They are major competitors in most energy markets. Gas plays an important role in producing electricity. Many electric utilities are also in the gas business. Gas companies are increasingly involved with cogeneration, an efficent technology for producing electricity. Gas is considerably cheaper than electricity in many applications. Both electric and gas utilities would benefit from greater use of gas during summer months, to reduce excessive demand for electricity for air conditioning.

Chapter 24. From the wellhead to the point of end use, natural gas enjoys significant economic advantages over oil. U.S. natural gas resources subject to economic recovery are more plentiful than those for oil. The primary recovery rate for natural gas wells is 70-80 percent, compared with 15-25 percent for oil wells. Natural gas can be utilized essentially the way it comes out of the ground, while oil has to be refined into usable products. For the consumer, natural gas is always available at the turn of a switch. Payment is due only after the gas has been used. In contrast, oil has to be ordered and paid for in advance of use. Oil also involves storage on the premises of the user. Gas burns cleaner than oil, which reduces air pollution and equipment maintenance expenses. Because of these advantages, gas is the preferred fuel for most stationary applications. Once the infrastructure of filling stations is in place, natural gas will also be able to compete successfully with oil-derived fuels in transportation markets.

Chapter 25. Between 1970 and 1987, world gas consumption outside the U.S.A. more than tripled, from 14.2 Tcf to 45.4 Tcf. The oil price escalations and supply disruptions of the 1970's played a major role in this upsurge of gas consumption. The environmental advantages of gas have given this fuel additional support in the marketplace. The Soviet Union is the world's largest producer, consumer, and exporter of natural gas. The U.S.A., which had pioneered the use of natural gas since the 1820's and was the world's leader until the 1970's, now is in second place in both gas production and consumption. All the industrial countries of the world have become major gas consumers. Leading gas producers include the USSR, U.S.A., Canada, Netherlands, United Kingdom, Algeria, Indonesia, Norway, and Mexico. The growing role of natural gas in world energy markets benefits consumers everywhere. It improves energy security and the environment. It also accelerates the introduction of new gas technologies, equipment, and appliances.

Chapter 26. The following major trends favor the increased use of natural gas in the foreseeable future: (1) Positive outlook for gas supply; (2) Reasonable price prospects; (3) Environmental advantages; (4) New technologies and applications, notably cogeneration, combined cycle, gas cooling, and natural gas vehicles; (5) Comparative advantages over competing energy sources; and (6) The worldwide growth of gas use and production.

Chapter 27. The following table ranks major energy sources in terms of important criteria. Each factor is ranked in a range from 1 to 10. The more desirable the rating, the higher the number.

Ratings of Energy Sources

Criteria	Gas	Oil	Coal	Elect.
Production costs	8	5	5	2
Transportation costs	6	8	6	8
Environmental costs	10	4	1	3
Consumer values	10	5	7	7
End use convenience	10	7	1	10
Appliances and equipment	7	7	7	10
Versatility	10	8	3	7
Energy security	10	3	9	9
Public image	10	3	4	7
Total	81	50	43	63

When all factors are added up, gas ranks first, with 81 points out of 90 possible. Gas is 18 points higher than electricity, 31 points above oil, and 38 points ahead of coal. It is the only energy source with six perfect scores. These high ratings support the conclusion that natural gas is the best energy choice.

1. The Favorable Outlook for Natural Gas

For most of the twentieth century, oil has been the dominant fuel in the United States and the rest of the world. The events of the past sixteen years, including unprecedented price escalations and supply disruptions, have eroded confidence in oil and have laid the foundation for the emergence of new energy leadership. On the basis of merit, natural gas is well suited for that role.

Natural gas has many advantages. Chemically, it is closely related to oil and can replace its liquid counterpart in most applications. It is available in plentiful amounts from domestic sources. It is the cleanest fossil fuel. A network of 1,150,000 miles of pipelines and gas mains is in place to serve the American people. Gas is convenient and user-friendly. Consumers have a wide choice of gas appliances and equipment. Taking all factors into consideration, gas is the best energy value for many applications.

Virtually any energy function performed by oil can also be handled by natural gas. In fact, natural gas often does the job better than oil. This comment is particularly applicable to space heating and hot water production, in which gas equipment is more efficient, requires less cleaning and maintenance, and is more user-friendly than oil equipment. In its compressed form, natural gas is an excellent vehicular fuel. It burns much more cleanly than gasoline or diesel fuel. The more natural gas is used to replace oil, the cleaner the environment will be.

The known gas resources in the United States that can be

economically recovered are far greater than the oil resources. Natural gas production has significant economic advantages over oil production. Because natural gas comes readily to the surface on its own, primary recovery from gas wells is 70 to 80 percent, while only 15-25 percent of the oil in place is obtained in the primary recovery phase. Additional oil production requires costly procedures. Natural gas can be utilized essentially in the form in which it comes out of the ground. Oil has to be refined into usable products, which adds to the cost.

More than 90 percent of the natural gas consumed in the U.S. comes from domestic sources; the rest is imported from Canada. In contrast, imports account for over 40 percent of oil requirements. Dependence on oil imports is growing as domestic resources are being depleted. The increased use of natural gas will strengthen energy security. Fortunately, oil and natural gas are produced by essentially the same enterprises. The shift from oil to gas will involve minimal economic disruption. On balance, the energy-producing states will benefit from increased use of natural gas.

From an environmental standpoint, natural gas is an ideal fuel. Impurities are removed *before* natural gas is allowed into the pipelines. The burning of methane (the principal ingredient of natural gas) involves its combination with oxygen. The end products of this process are water vapor and carbon dioxide. The chemical formula for this transaction is CH_4 (methane) plus $2O_2$ (oxygen) equals $2H_2O$ (water) plus CO_2 (carbon dioxide). If it were not for the presence of nitrogen in the air, the combustion of methane would be as benign as human breathing, which also takes in oxygen and gives out water vapor and carbon dioxide. In the presence of an open flame, nitrogen combines with oxygen and forms compounds that are pollutants. However, even this negative feature is lessened with natural gas in comparison with oil and coal. While this phenomenon has not been fully explained, a possible reason is that methane has such a strong affinity for oxygen that it reduces the availability of this element for combining with nitrogen.

Methane burns so efficiently and cleanly that it can help reduce the pollution caused by the combustion of coal. Methane has the qualifications to be used as a pollution control procedure in relation to coal. In this application, it is more economical than alternative technologies.

A network of 1,150,000 miles of gas pipelines and mains serves more than 43 million American homes, over 4 million commercial customers, and about 200,000 industrial users. An estimated 172 million Americans are served by the gas industry. By turning on a switch, or with an automatic ignition device, the home owner can use gas to cook a meal, heat the house, produce hot water, and dry the laundry. New efficient gas cooling equipment will also be available for homes in the near future. In commercial buildings, gas can fulfill all the functions cited for residential applications. In addition, gas is finding a growing market for commercial air conditioning. Cogeneration, a technology that generates both electricity and usable heat, is of particular interest to many commercial users, including hospitals, schools, office buildings, hotels, and apartment houses. Packaged gas-fired cogeneration systems are now available with a wide range of capacities.

Natural gas has many industrial applications. The use of gas is compatible with industry's efforts to achieve high levels of energy efficiency at reasonable cost.

Gas equipment manufacturers offer a great variety of high efficiency furnaces and boilers. Many models achieve efficiencies of 90 percent or more. Similarly, highly efficient gas water heaters are now available. Kitchen chefs and cooks prefer gas equipment because is allows them maximum control over the flame. A wide range of gas-fired kitchen equipment is available for homes as well as for commercial kitchens.

The increased use of natural gas can save consumers large amounts of money. Some eight million households currently connected to gas could save an estimated $2 billion or more annually by making hot water with gas rather than with electricity or oil. Gas water heaters are relatively inexpensive. The return on such an investment can be as much as 40-60 percent annually. Very large savings can also be achieved by switching from oil or electricity to gas for space heating. Similarly, new gas cooling equipment can generate big savings on electric bills. For commercial enterprises, cogeneration can be a big money saver. Altogether, the more than 52 million residential and commercial users of natural gas can save many billions of dollars a year by making greater use of this fuel. Those who are not yet connected to gas but have the opportunity to gain access to this fuel, should take advantage of its many cost-saving options.

Natural gas vehicles are likely to find increased acceptance among fleet operators. Such vehicles have substantial environmental advantages over their counterparts powered with gasoline or diesel engines. In addition, they can produce significant savings on fuel costs. Buses, trucks, and utility vehicles with dedicated natural gas engines are being introduced in the marketplace. They can play an important role in cleaning the air in urban environments now suffering from smog.

On the basis of its cleanliness, efficiency, versatility, user-friendliness, attractive value, and availability from domestic sources, natural gas is the best energy choice for many applications.

Throughout this book, the terms "natural gas" and "methane" are used interchangeably. Natural gas which is transported by pipelines consists primarily of methane, a hydrocarbon molecule made up of one carbon atom and four hydrogen atoms (CH_4). This molecule is the subject matter of the book.

PART I
Methane Gas Resources and Supply

2. The Origins of Methane

Methane, the principal ingredient of natural gas, can be found near the surface of the earth as well as in deep locations. Oil fields and coal seams contain methane. Organic debris can be converted to methane by the action of microorganisms. Wherever methane is found, it has the same basic composition, namely one carbon atom combined with four hydrogen atoms (CH_4).

There is broad agreement among geologists that methane originates from the following sources: (1) Organic matter in sediments whose decomposition is being promoted by heat; (2) The action of anaerobic microorganisms that convert organic materials into methane; (3) The transformation of oil and other heavy hydrocarbons into methane at high temperatures, usually in deep locations; and (4) Coal, which releases methane as it matures.

It is noteworthy that all of these sources of methane are biological in nature. They originate with plant materials that have been transformed into methane. Because plants derive their energy from the sun, it is appropriate to credit solar energy as the ultimate source of methane.

Some scientists have presented the hypothesis that methane was part of the earth's original makeup.[1] This hypothesis states that methane is widely present in the universe and was included in the earth's primordial mass. Moreover, methane can be formed by the interaction of other originally present gases, notably carbon dioxide and hydrogen. Much of this methane may be situated in the earth's mantle, the section between the crust and the core. Under certain geological conditions, this methane moves closer to the surface. Insofar as methane comes from this source, it is *abiogenic* (not of biological origin). This hypothesis remains to be proven and is not generally accepted by geologists.

Methane has been described as "the gaseous phase of petroleum."[2] It is likely to be found wherever petroleum is being formed. Organic matter in sediments decomposes under the impact of heat. Among the end products of this decomposition are a variety of hydrocarbons, including methane. While some methane forms along with the liquid petroleum, the main creation of methane occurs at a temperature higher than that best suited for oil. The formation of oil in sediments tends to peak around 100° Celsius (the boiling point of water); the bulk of methane generation occurs at 150° Celsius.[3]

The production of methane by microorganisms can be observed at landfills. When garbage is covered to exclude air, methane soon begins to form. The producers of this methane are living organisms that can function only in an oxygen-free environment. These minute forms of life have been given the scientific name anaerobic ("no air") microorganisms. The same process that converts decaying organic garbage into methane takes place on a vast scale throughout nature. Whenever organic debris is removed from contact with the air, the methane-forming microorganisms are ready to do their work. This process plays an important role in keeping the earth a habitable planet. If it were not for the work of these anaerobic microorganisms, the earth would have been buried in organic debris a long time ago. Anaerobic microorganisms may also be the key to providing mankind with renewable energy from methane. They are ready to perform their function of converting biomass into methane whenever human beings decide to make use of this resource. It is possible that some time in the future methane will be produced in giant energy factories consisting of biodigesters in which anaerobic microorganisms work ceaselessly at their task of ingesting organic materials and giving off methane.

Methane is the most stable of all hydrocarbons. It can exist unchanged at temperatures as high as 550° Celsius (1022° Fahrenheit). All other hydrocarbons, including oil, tend to be converted into methane as temperatures increase. This process takes place in oil reservoirs at deeper locations, where the temperatures of the earth are higher. Under these conditions, the earth acts like a giant refinery, which converts heavy liquid hydrocarbons into light gaseous forms, with methane the ultimate product. Another factor that contributes to the conversion from oil to methane is the geologic age of the petroleum. The longer the oil has been in existence, the more it is likely to be converted into methane.[2,3]

The large quantities of methane that are found in some deep locations (15,000 or more feet below the surface) may have resulted from formerly giant oil fields that migrated from shallow to deep locations or that were pushed into great depths by geological upheavals. Because of the higher temperatures prevailing at greater depths, the oil was transformed into methane. Giant deep gas fields may be the residuals of formerly giant oil fields.

Coal tends to generate methane as it matures. Large quantities of methane are locked up in coal seams. Recent technological advances have made some of this methane accessible to economic utilization. The removal of methane from coal seams makes coal mining safer and increases the useful energy available for human needs.

Some scientists have presented the hypothesis that methane may have been part of the earth's original makeup. Because this hypothesis focuses on the non-biological environment, it has been labeled the "abiogenic" theory of methane's origin. This hypothesis may be summarized as follows:[4]

(1) Ever sinces the earth was formed, large quantities of methane have been present in the planet's interior, with most of it being trapped in the mantle and some of it having migrated to the top 25 miles of the earth's crust.

(2) Other gases which are present in the earth's mantle, such as carbon dioxide and hydrogen, may combine to generate additional methane.

(3) Methane from deep in the earth travels to the crust through openings created by fractures and by volcanic or earthquake activities.

(4) Abiogenic methane has been trapped by rock, sand, and salt formations in the crust of the earth.

The abiogenic hypothesis has been championed by Professor Thomas Gold, an astrophysicist at Cornell University.[1] From a practical point of view, the verification of the abiogenic hypothesis would add immensely to the potential methane resources. Moreover, it could lead to exploration for methane gas in areas previously not considered good prospects, such as deep faults in the earth, volcanic areas, and earthquake zones.

Even without the substantiation of abiogenic methane, the natural gas resources of the earth are enormous. The biological sources of methane should be adequate to take care of mankind's energy needs for a long time to come. Moreover, if scientists and energy decision makers move forward with the development of technologies for the economic utilization of anaerobic microorganisms for methane production, gas energy would be available in adequate quantities for the rest of mankind's existence.

Sources:

[1] *"Terrestrial Sources of Carbon and Earthquake Outgassing," by Thomas Gold,* Journal of Petroleum Geology, *Vol. 1, No. 3, 1979.*

"Studies Related to the Deep Earth Gas," by Thomas Gold, Elizabeth Bilson, and Steven Soter, Center for Radiophysics and Space Research, Cornell University, prepared for the Gas Research Institute, May 8, 1981.

"Abiogenic Methane and the Origin of Petroleum," by Thomas Gold and Steven Soter, Cornell University, October 15, 1981.

[2] Petroleum Geochemistry and Geology, *by John M. Hunt. W.H. Freeman and Co., San Francisco, 1979.*

[3] *"The Future of Deep Conventional Gas," by John M. Hunt. UNITAR Conference on Long-Term Energy Sources, Montreal, December 1979. Published by Pitman Publishing, Inc., Marshfield, MA 02050, 1981.*

[4] *"Abiogenic Methane: A Review,"* Gas Energy Review, *American Gas Association, May 1980.*

3. Methane Gas Energy Forever!

Methane, the principal ingredient of natural gas, consists of one carbon atom combined with four hydrogen atoms (CH_4). This molecule is potentially our most plentiful energy resource. It is present in large quantities in many parts of the United States, in depths ranging from just below the surface to more than 30,000 feet. Sizeable amounts are locked up in tight formations, coal seams, brine, and frozen hydrates. Methane can be obtained from gasification of coal, peat, and oil shale. It can be manufactured from organic wastes and from biomass. The state of Alaska has huge natural gas resources, which have only begun to be tapped. Altogether, the methane gas resource base in the United States is large enough to justify an optimistic outlook for the future. Methane will play an important role in providing energy as long as human beings live in the United States.

The following table summarizes estimated methane gas resources in the United States. At the current rate of consumption of approximately 19 Tcf per year, the conventional gas resource base of 983 Tcf would take care of requirements for fifty years. Other estimated methane sources in tight formations, totaling 2,800 Tcf, would be adequate for 150 times current usage. Gasification of coal, peat, and oil shale could provide enough methane to last for a thousand years. Methane from biomass is renewable. As long as the sun shines and plants grow, human beings can manufacture all the methane gas they need for the rest of time.

Estimated U.S. Methane Gas Resources

Resource Category		Trillion Cubic Feet
Conventional resources[1]		
Proved resources		187
Probable resources (current fields)		164
Possible resources (new fields)		238
Speculative resources (frontier areas)		185
Coalbed methane		90
Alaska		119
	Total	983
Other methane sources in tight formations[2]		
Tight sands		600
Devonian shale		200
Coal seams		800
Geopressured in brine		1,200
	Total	2,800
Nonconventional sources of methane[2]		
Coal gasification		10,000
Peat gasification		1,440
Oil shale gasification		7,860
	Total	19,300
Methane from biomass[3]		renewable
Landfills		
Animal wastes (manure)		
Sewage		
Energy Crops		

The utilization of these methane gas resources is primarily a function of technology and economic cost. Generally speaking, conventional gas resources are produced with known technologies at the lowest cost. However, some of the conventional resources, notably those in locations 15,000 or more feet below the surface, and offshore resources, involve complex technologies and high costs, which are justified if sufficient quantities of gas are found. While conventional gas resources in Alaska are very large, their utilization awaits the construction of pipelines to supply U.S. markets.

Some methane from tight formations is currently being produced. It generally comes from the more favorably situated deposits. Additional technological developments are necessary to make greater use of these resources at competitive prices. Similar comments are applicable to nonconventional sources of methane.

Methane from biomass is a renewable resource. While the cost of producting methane from plants and other organic materials is higher than conventional gas, it can compete successfully in a number of applications. For example, methane from landfills has the advantage of being located in the service territories of gas companies, which eliminates long-distance transportation costs.

It is important to view the methane gas resource base as a dynamic foundation for meeting future energy requirements. The conventional gas resources estimates presented in the preceding table may prove to be too conservative, particularly in relation to deep gas, offshore gas, and Alaskan gas. As technology progresses and/or economic conditions change, additional methane gas may become available at costs that compare favorably with conventional gas. For example, until fairly recently all methane from coal seams was considered too costly and too difficult to recover for inclusion among conventional resources. However, a combination of technical advances and federal tax incentives has made 90 Tcf of coal seam methane suitable for upgrading to the status of a conventional resource.

As the price of conventional gas rises, other methane resources will become more competitive. Methane gas resources that are hardly utilized currently may become major contributors to meeting energy needs in the future. As demand for methane increases, additional supplies will be forthcoming. The U.S. is fortunate in having such a large and multifarious methane gas resource base that provides a sound foundation for its energy requirements.

Sources:

[1] *"New Report Reaffirms Magnitude of Nation's Sizeable Natural Gas Resource," Potential Gas Agency, Colorado School of Mines, Golden, CO 80401, February 22, 1989.*

[2] New Technologies for Gas Energy Supply and Efficient Use, 1983 Update, *American Gas Association, May 1983.*

[3] The Gas Energy Supply Outlook 1987-2010, *American Gas Association, October 1987.*

4. The Large Conventional Gas Resource Base

The U.S is well positioned to enter the gas age. Plentiful conventional gas resources provide the foundation for a secure energy future that will also bring significant environmental benefits.

Conventional gas resources may be subdivided into proved reserves, onshore gas in shallow and deep locations, offshore gas, and Alaskan gas. The resource estimates cited in the preceding chapter are conservative. Some experts believe that these resources may be much larger, particularly gas in deep locations (below 15,000 feet) and in offshore areas.

Proved Reserves, the Gas Industry's Inventory

Proved reserves are natural gas resources which have been discovered, measured to a high degree of accuracy, and which can be readied for delivery to customers in a relatively short period of time. In 1988, proved reserrves totaled 187 Tcf, which was about ten times consumption. This ratio was normal and adequate to take care of requirements. Proved reserves may be considered the inventory of the gas industry.

When customers purchase gas, it is withdrawn from the proved reserves. Gas producers obtain the funds with which to do additional drilling to replenish the gas that has been sold. Gas demand is the prime driving force that induces gas production and the creation of proved reserves. If demand increases, proved reserves will grow as well. The natural gas resource base is sufficiently large to make possible adequate proved reserve levels for many years to come.

Lower 48 States Conventional Onshore Gas Resources

Conventional onshore gas resources in the lower 48 states have been estimated at 428 Tcf, of which 283 Tcf are in shallow locations and 145 Tcf are situated more than 15,000 feet below the surface. Most of this gas is located in the Mid-Continent (139 Tcf), Rocky Mountains (120 Tcf), and Gulf Coast (75 Tcf).[1] Most of the offshore gas is also accounted for by the Gulf Coast (129 Tcf).[1] In 1987, total gas production in the U.S. amounted to 16,114 Bcf. Texas led the way with 4,517 Bcf, followed closely by the federal government's offshore holdings (4,462 Bcf). The other leading producers were Oklahoma (1,813 Bcf) and Louisiana (1,625 Bcf).[2]

At year-end 1988, the following companies had gas reserves in excess of one trillion cubic feet in the continental U.S.: Amoco, 12.4; Exxon, 9.3; Chevron, 8.3; Mobil, 7.5; Arco, 6.4; Shell, 6.2; Texaco, 4.7; Unocal, 4.3; Burlington Resources, 3.8; Phillips, 3.4; USX, 3.0; Occidental, 2.6; BP America, 2.3; Mesa, 2.3; Sun, 2.2; Dupont, 2.2; Anadarko, 1.7; Union Pacific, 1.4; Enron, 1.2; Consolidated Natural Gas, 1.2; Coastal, 1.0.[3] Many other companies own substantial gas reserves. At the present time, no one company or small group of companies dominates the gas supply field. There is healthy competition among gas producers.

Conventional gas from shallow wells (less than 15,000 feet) has been produced in the U.S. for more than 150 years. The technologies involved in shallow gas production are well known. Many companies and individuals have the know-how and financial resources to drill such wells. While substantial amounts of gas in shallow locations remain to be discovered in the continental U.S., it is likely that the 283 Tcf estimated by the Potential Gas Committee is close to the mark. In contrast, the estimates for deep gas (145 Tcf) may be on the low side, This topic will be more fully explored in the next section.

Deep Gas, A Frontier with Great Promise

Methane, being the most stable of all hydrocarbons, can exist at temperatures and pressures much greater than those suitable for oil. The physical characteristics of methane remain intact at temperatures as high as 1,022° Fahrenheit.[4] In contrast, oil starts to break down when temperatures reach 450° Fahrenheit. The deeper the location, the higher the temperatures and the greater the pressures. These physical realities favor the discovery of methane gas over oil in deep drilling.

It is noteworthy that deep gas was not considered a viable commercial source of conventional gas until the last decade. Technological breakthroughs achieved in seismic exploration, drilling pipe, and proppants made possible the increased drilling for deep gas. Moreover, the price of gas had to rise before the large amounts of capital required for drilling deep wells could be justified. While several thousand deep wells have been drilled, the resource base is not nearly as well known as the one for shallow gas.

The Potential Gas Committee has estimated methane gas resources below 15,000 feet at 145 Tcf. Some independent gas producers are considerably more optimistic about the prospects for finding large quantities of gas at great depths. For example, Mr. Robert Hefner III, Managing Partner of GHK Corporation, has estimated that the deeper layers of the Anadarko Basin in Oklahoma and Texas may contain between 60 and 300 Tcf of gas.[5] Similarly, Mr. Philip F. Anschutz, President of Anschutz Corporation, expressed the view that the gas resources in the Rocky Mountain Overthrust Belt may total more than 200 Tcf.[6] Mr. L.W. Funkhouser, a leading expert on deep gas geology with Chevron, has estimated the gas resources of the Tuscaloosa Trend in southern Louisiana at 60 Tcf.[7] This formation covers an area about thirty miles wide and two hundred miles long. Its energy content is estimated to be the equivalent of a ten billion barrel oil field.

Support for such optimistic appraisals of deep gas resources is provided by Dr. John M. Hunt, a petroleum geologist on the staff of the Woods Hole Oceanographic Institution. He pointed out that because of the physical characteristics of methane, it may be the only hydrocarbon found in the deepest and oldest geological sediments. Dr. Hunt noted that these deep gas fields may be partly the result of giant oil fields that moved to lower depths during geological upheavals. Under those conditions, the pressures and temperatures in the earth converted the oil into methane. Dr. Hunt pointed out that the Tuscaloosa Trend may be characteristic of other geological formations, where former giant oil fields have been buried at great depths.[4,8]

Several technological developments were essential before deep gas drilling became commercially feasible. Advanced seismic technologies and specialized computer programs were designed to improve the accuracy of discovering the presence of hydrocarbons at great depths. As a result, the success rate of finding gas in deep locations has improved in relation to discovering gas in shallow areas.

The pipe that is used for deep drilling has to be made of special steel alloys that can withstand the enormous pressures, high temperatures, and corrosive substances that are encountered at great depths. Such drilling pipe is now readily available.

Once a hole is drilled, it is essential to keep it open for the gas to flow. The openings in gas wells have a tendency to close up because of pressures from the surrounding areas. If such an occurrence were to take place at a shallow location, coarse-grained sand would be used as a proppant to keep the hole open. However, at great depth sand is crushed to a fine powder by the pressures. To solve this problem with deep wells, Exxon developed a special proppant made of sintered bauxite. This proppant has proved successful in keeping the fractures open and the gas flowing from deep wells.[9]

Drilling for deep gas is very expensive. Mr. Funkhouser of Chevron noted that development wells in the Tuscaloosa Trend range in cost from $6 million to $15 million.[7] These figures compare with about $200,000 average cost for shallow wells.[10]

Inspite of the high initial costs and risks associated with drilling for deep gas, successful ventures can bring substantial rewards. The following factors make deep gas drilling commercially feasible for companies with large financial resources and experience in this specialty:

(1) A deep gas well can be drilled, completed, and put into production in a relatively short time, often no more than a year. As a result, such wells can generate income fairly quickly.

(2) The economics of successful deep gas drilling require the presence of significant amounts of gas. If a large gas formation is tapped, the chances for a profitable venture are good. Under favorable conditions, a deep gas well may recover gas for many years.

(3) Successful deep gas wells produce much more gas than shallow wells. Many deep gas wells produce several million cubic feet of gas a day. Such prolific production helps pay for the high costs involved and can soon lead to substantial proftis.

(4) Primary recovery of the gas in place in deep wells averages 70 to 80 percent. Methane, which is lighter than air and under pressure at great depth, readily comes to the surface by itself.

(5) Many of the deep gas wells that have been drilled thus far are situated near existing gas pipelines. As a result, gas from such fields can be brought to market quickly, without necessitating costly pipeline construction. These comments are applicable to deep gas fields in the Tuscaloosa Trend and in the Anadarko Basin.

Deep gas has the potential for becoming one of the most important gas frontiers. Dr. Hunt expressed the view that "Deep conventional gas accumulations may well be our major fossil fuel...when the oil begins to run out sometime in the next century."[8]

Offshore Gas, Another Promising Resource

The Potential Gas Committee has estimated offshore gas

resources in the continental U.S. at 159 Tcf. Major technological developments have greatly expanded the scope of exploration and production from offshore gas resources. Drilling vessels and platforms now drill wells in water depths of 7,000 feet or more. It is technically feasible to install subsea wellheads in 7,500 feet of water and connect them by pipeline to production facilities.[11]

Satellite navigation has greatly improved seismic exploration for offshore gas. Sophisticated computer programs and seismic technologies are providing increasingly reliable information concerning gas resources covered by large amounts of water.[11]

Offshore gas exploration and production involves specialized skills and large amounts of capital. As these factors are brought to bear on the U.S. offshore areas at ever deeper water locations, the methane gas resources are likely to grow significantly beyond the 159 Tcf estimated by the Potential Gas Committee.

Alaskan Gas, a Major Resource for the Future

Alaskan natural gas resources have been estimated at 152 Tcf, of which 33 Tcf are proved reserves.[11] Most of the latter are situated in the Prudhoe Bay field on Alaska's North Shore. This gas has not yet been utilized because it lacks a means of transport for bringing it to market. Current gas production in Alaska is confined to Cook Inlet in the southern part of the state. In 1986, Alaska produced 324 billion cubic feet (Bcf) of gas. About 50 Bcf were exported to Japan in the form of liquefied natural gas (LNG). The rest was used for local utility operations, for oil production, or for direct sale to end users.[11]

To bring Alaskan gas to the lower U.S. involves the construction of a 4,800 mile pipeline from Prudhoe Bay through Canada to the U.S. This project, called the Alaska Natural Gas Transportation System (ANGTS) was approved by the U.S. Congress in 1976. When completed, this pipeline could deliver two Bcf of gas daily, which could later be expanded to 3.2 Bcf daily. Phase 1 of this pipeline, which originates in Alberta, Canada, has been completed. It carries Canadian gas to the

Midwest and the West Coast. Plans to extend this pipeline to the Mackenzie Delta and the Arctic Delta in northwestern Canada are currently in an active phase. (See chapter on Canadian gas imports for additional information.) Once this segment is approved and constructed, which will take several years, an extension to the Prudhoe Bay field will be required. It will probably be some time after the year 2000 before Alaskan gas reaches the continental U.S. via this route.

The geological formations for conventional gas are very favorable in Alaska. Once pipelines are constructed, Alaska will become a major source of natural gas for the U.S. in the twenty-first century.

A Look at the Future of Conventional Gas

According to the estimates of the Potential Gas Committee, conventional gas resources of the U.S. were 983 Tcf at the end of 1988, enough to last about fifty years at current rates of consumption. Does this mean that by the year 2038 we will have run out of conventional methane gas? Fortunately, the outlook is more favorable than might be deduced from a mechanistic application of arithmetic. It should be remembered that the Potential Gas Committee estimates are based on conservative assumptions about resources and technologies. They try to keep speculation to a minimum, which is an appropriate procedure for their purposes.

However, one can make a reasonable case for assuming that the conventional gas resources available in the year 2038 may be as large as those in 1988, if not larger. The explanation for this apparent paradox revolves around several factors:

(1) Over the next fifty years it is likely that additional gas resources will be discovered below 15,000 feet that are not currently included in estimates. A new frontier may open up in the range of 25,000-50,000 feet, which might tap hitherto unknown gas resources. If large amounts of abiogenic gas exist deep in the earth, they are more likely to be reached through deeper drilling.

(2) Drilling in deep ocean locations will also be extended to new frontiers, with possible prospects of finding gas reosurces not currently included in estimates.

(3) The technologies for recovering methane from tight formations and nonconventional sources are likely to advance sufficiently to make these resources available at costs comparable to those applicable to conventional resources. The likelihood of such a development is increased in view of the reality that convential gas prices will probably rise with the passage of time. Such price increases need not be a cause for alarm, particularly if they are gradual and if improvements in efficiency of gas appliances and equipment keep pace with price increases.

In regard to long-range forecasts for conventional natural gas resources, optimism is more realistic than pessimism.

Sources:

[1] *Estimated conventional natural gas resources were derived from data supplied by Robert Kalisch, Director, Gas Supply and Statistics, American Gas Association. Another source was a news release from the Potential Gas Agency, Colorado School of Mines, Golden, Colorado, dated February 22, 1989.*

[2] *"Estimated Total Dry Natural Gas Proved Reserves, Reserves Changes, and Production, 1987,"* U.S. Crude Oil, Natural Gas, and Natural Gas Liquids Reserves, 1987, *Energy Information Administration, 1988.*

[3] *"Preliminary Findings Concerning 1988 Natural Gas Reserves," American Gas Association, April 28, 1989.*

[4] *"The Future of Deep Conventional Gas," John J. Hunt,* UNITAR Conference on Long-Term Energy Sources, *December 1979, published by Pitman Publishing Inc., Marshfield, MA, in 1981.*

[5] *"Deep Drilling in the Anadarko Basin," by Robert A. Hefner III,* Petroleum Engineer International, *March 1979.*

[6] *"A Review of the Overthrust Belt," by Philip F. Anschutz, address to the American Gas Association, October 22, 1979.*

[7] *"The Deep Tuscaloosa Gas Trend of South Louisiana," by L.W. Funkhouser, F.X. Bland, and C.C. Humphris, Jr., paper presented at the national meeting of the American Association of Petroleum Geologists, June 8, 1980.*

[8] Petroleum Biochemistry and Geology, *by John M. Hunt, published by W.H. Freeman & Co., San Francisco, 1979.*

[9] *"Use of High-Strength Proppant for Fracturing Deep Wells," Claude E. Cooke, Jr., John L. Gidley, and Dean J. Mutti, Exxon Company. This paper was presented at the 1977 Deep Drilling and Production Symposium of the Society of Petroleum Engineers, held in Amarillo, Texas, April 17-19, 1977.*

[10] *"Primer on Natural Gas and Methane," Scientists Institute of Public Information, November 15, 1979.*

[11] The Gas Energy Supply Outlook 1987-2010, *American Gas Association, October 1987.*

5. Tight Formations Contain Large Treasures of Methane

Large quantities of methane gas are trapped in tight sands, Devonian shale, coal seams, and geopressured brine. The recovery of methane from these sources is a function of experience, technology, and price. It is likely that these resources will be increasingly utilized with the passage of time.

Recovery of methane from tight formations is not new. In fact, current production from tight formations in the continental U.S. is estimated at 1.4 Tcf, or almost nine percent of total gas production.[1] This gas comes from sources that are known and accessible with existing technologies. As technology advances and/or energy prices increase, additional production from tight formations can be achieved.

Tight Sands

Methane trapped in tight sand formations has been estimated at 954 Tcf.[2] These resources are situated primarily in the western U.S.

To stimulate gas production from tight sands, the formation holding the methane must be fractured. The most common procedure is to pump fluids (primarily water) into the formation. To keep the gas flowing, the well is kept open with proppants.

The Department of Energy (DOE) is carrying on research in the Piceance Basin of Colorado to improve understanding of tight sands. The Gas Research Institute (GRI) is working on improved technologies to increase production from tight sands.[3]

The following table estimates recoverable methane from tight sands at varying prices and levels of technology.

Recoverable Gas from Tight Sands[2]

Gas Price 1987 $/Mcf	**Base Technology** Tcf	**Advanced Technology** Tcf
2.50	84	366
5.00	178	542
9.00	241	584

This table shows that advanced technology is the most important factor in determining methane recovery from tight sands. The large resource base justifies increased efforts to develop such technologies.

Devonian Shale

The Appalachian region of the eastern U.S. has produced natural gas for about 130 years, making it one of the oldest gas producing areas of the country. Thousands of producing wells dot the area, but they produce only small quantities of gas. The average production is 79,000 cubic feet per day.[4] Many of these wells have been producing for decades. Current production totals about 400 Bcf annually.[5] The Appalachian Basin covers an area of about 160,000 square miles. It is underlain with Devonian shale, which is about 400 million years old. Methane gas resources of this region have been estimated to range form 225 Tcf to 1,800 Tcf.[1]

The traditional method of producing gas from Devonian shale has been to use explosives (mostly nitroglycerine) to fracture the stone and keep the gas flowing. However, this approach is limited to wells situated in shallow locations. It is estimated that using this technology, total resource recovery would be about 25 Tcf.[1]

Starting in the 1970's, hydraulic fracturing and other advanced technologies have been used to stimulate production from Devonian shale. It is estimated that the wider application of these technologies would double the recoverable methane.[1]

Research on methane from Devonian shale is carried on by gas companies situated in the area, as well as the Department of Energy and Gas Research Institute. The location of the Appalachian region near major gas consuming markets is an important advantage.

Devonian shale averages about 8,000 feet in depth. Drilling for gas in the same location, but at much greater depth, has yielded some important gas discoveries. For example, Columbia Gas drilled a well in Mineral County, West Virginia, which had an initial flow of 88 million cubic feet of gas a day, or about one thousand times the flow of wells from Devonian shale. Other deep gas discoveries below the Devonian shale have been made in Pennsylvania and New York.[6] This gas comes from geological formations different from Devonian shale. It remains to be seen whether there is any relationship between the deep gas and the gas trapped in Devonian shale.

Coal Seams

Methane trapped in coal seams in the U.S. is estimated at 800 Tcf of which about 100 Tcf may be recoverable.[1]

The outlook for recovering methane from coal seams has greatly improved in the recent past. Tax incentives (approximately 70 cents per Mcf) combined with advanced technology have provided the motivation for substantial investments in this resource by private companies. For example, Arco has found 500-600 Bcf of coal seam gas in the Arkoma Basin of Oklahoma in 1987-88. Burlington Resources added more than 800 Bcf to its reserves in 1988, nearly all from methane in coal seams.[7]

Current production of methane from coal seams is about 25 Bcf/year. This amount will at least double by the year 2000 and may increase substantially more if gas prices rise.[1]

If methane is left in coal seams, it poses a hazard to coal mining. Therefore, it is in the mutual interest of the coal mining companies and the gas industry to make economic use of the methane. However, synchronizing methane recovery and coal mining may pose problems. Moreover, many states have confusing statutes about the ownership of the methane in coal seams. As a result, utilization of this resource has been held back.[8]

The U.S. Bureau of Mines and the Department of Energy have carried on research concerning this resource. Considerable progress has been made in solving the technical problems involved in recovering methane from coal seams.[9]

Geopressured Methane in Brine

Vast amounts of methane are trapped in brine in the Gulf Coast area. Estimates of this resource run as high as 100,000 Tcf, but most of this methane is not economically recoverable. However, some of these resources are high quality deposits that have good prospects for commercial recovery.

The Texas-Louisiana Gulf area is one of the most prolific gas-bearing regions of the country. About 175 Tcf of gas has already been produced there, with an additional 120 Tcf gas production likely. A substantial number of gas-bearing formations in this area have been flooded with brine. This methane is concentrated in known locations. It can be recovered by removing the brine. While this procedure involves technical challenges and increases the cost of production, recovery of this resource is feasible, particularly at higher gas prices. The Institute of Gas Technology has estimated this high-grade geopressured methane in brine at 60 Tcf.[1]

Over the near term, production from this resource could reach 50-100 Bcf a year. If gas prices rise sufficiently, production of methane in brine could reach one Tcf or more by the year 2010.[1]

Because of the large resource potential, geopressured gas in brine has aroused considerable scientific interest. The

University of Texas at Austin and Louisiana State University have been leaders in this field.[10]

Methane Hydrates

Methane hydrates are solid, ice-like compounds in which methane is entrapped and bound to water molecules. Permafrost zones and cold ocean sediments are the most likely places where gas hydrates may be found. Permafrost covers about 23 percent of the earth's landmass, including 47 percent of the Soviet Union, 75 percent of Alaska, and 63 percent of Canada. Permafrost depths may exceed 2,000 feet. It is believed that onshore hydrate-bearing zones were originally natural gas that became frozen over geologic time due to Arctic conditions.

Estimates of this resource are highly speculative, but the amounts of gas involved are considered to be enormous. Some estimates run as high as 6.7 million Tcf in the U.S. and 270 million Tcf worldwide.[1]

Scientists in the Soviet Union have done considerable research on this resource. In the U.S., research on methane hydrates has been conducted by the National Science Foundation and the Office of Naval Research.[11]

Major technological breakthroughs are required before this resource can be commercially utilized. It will probably be well into the twenty-first century before such technologies are developed.

Methane gas from tight formations will play an increasingly important role in supplying U.S. energy requirements in the future. The vast resources justify additional research to develop technologies that will make this methane available at reasonable prices.

Sources:

[1]The Gas Energy Supply Outlook 1987-2010, *American Gas Association, October 1987.*

[2]*"Tight Gas Analysis System," by Lewin and Associates, Inc., March 1987.*

[3]*"Tight Gas Technology Status and Programs," Appendix D of* The Gas Energy Supply Outlook 1987-2010.

[4]*"Oil and Gas Potential of the Appalachian Thrust Belt," by Porter Brown, presented to the Potential Gas Committee, Scottsdale, Arizona, October 19, 1979.*

[5]*"Gas from Devonian Shale—Status and Outlook,"* Gas Supply Review, *American Gas Association, May 1978.*

[6]*"New Finds Heat Appalachian Basin Interest,"* Oil & Gas Journal, *February 11, 1980.*

[7]*"Preliminary Findings Concerning 1988 Natural Gas Reserves,"* Energy Analysis, *American Gas Association, April 28, 1989.*

[8]*"Methane from Degasification of Coalbeds,"* Gas Supply Review, *American Gas Association, June 1977.*

[9]*"Natural Gas from Coalbeds," by Troyt B. York,* Long-Term Energy Resources, *Pitman Publishing Inc., Marshfield, MA, 1981.*

[10]New Technologies for Gas Energy Supply and Efficient Use, *American Gas Association, April 1979.*

[11]*"Gas from Natural Gas Hydrates,"* Gas Energy Review, *American Gas Association, August 1979.*

6. One Thousand Years of Methane from Solid Fossil Fuels

A large part of the organic residues from life on earth throughout time is contained is such fossil fuels as coal, peat, and oil shale. The quantities involved are enormous. If U.S. coal resources were gasified, they could provide sufficient methane gas to last at least five hundred years at current rates of consumption. Peat gasification could yield an estimated 1,440 Tcf of methane, enough to take care of requirements for more than seventy years. Oil shale resources are comparable to those of coal.

The solid fossil fuels contain a complex range of substances that represent the residues from the original biological sources. One can think of them as storehouses of organic and inorganic chemicals. When these fossil fuels are burned, the various constituents combine with oxygen and/or with each other to form compounds that enter the environment. This process is a major source of air pollution.

Burning solid fossil fuels wastes valuable resources and pollutes the environment. The best procedure for avoiding these problems is to gasify the fuels. Gasification separates the energy portion from the rest. The energy part is made up of carbon and hydrogen, which is converted into methane. The remaining substances can be used as sources for chemicals and construction materials. It should be noted that chemicals are pollutants only it they are in the wrong place. For example, sulfur is an essential ingredient of many important chemical compounds;

it becomes a pollutant only if it is spewed into the atmosphere in the form of sulfur dioxide.

One procedure for gasifying fossil fuels is to mine them and process them through a series of physical and chemical steps that yield methane gas and various byproducts. The other alternative is to burn them in the place in which they are situated by adding oxygen. This procedure is called "insitu gasification." The methane gas that is formed by either process tends to have a lower energy content than pipeline quality natural gas, which averages 1,000 Btu (British thermal units) per cubic foot. The methane from gasification can either by upgraded to pipeline standards, or it can be used with lower energy content fur such purposes as the generation of electricity or to power industrial facilities.

The advantages of gasification may be summarized as follows:

(1) The resources involved are so large that they would assure the availability of methane gas for one thousand years of more.

(2) Almost all types of coal, peat, and oil shale can be gasified. This comment is particularly important in connection with fossil fuels that cannot be economically recovered or utilized with existing technologies.

(3) Methane from gasification has access to over one million miles of pipelines and more than 50 million customers in the United States. The pipelines represent an investment of $90 billion (and a much higher replacement cost). Moreover, users of natural gas have invested about $90 billion in gas appliances, furnaces, and other consumer equipments.[1]

(4) Methane gas is the fuel of choice for such applications as space heating, hot water production, cooking, and clothes drying.

(5) In the future, methane gas from gasification of fossil fuels is likely to play an increasingly important role in the production of electricity. For this purpose, the methane does not need to achieve pipeline energy levels. Moreover, electric power generating facilities can be situated near the gasification projects, which has economic advantages. Similar comments

are applicable to industrial facilities that use large amounts of mechanical and/or electrical power.

(6) Fossil fuels used for gasification are often situated near major consuming markets. As a result, transportation costs for the methane are significantly reduced.

(7) Gasification of fossil fuels is environmentally far more benign than burning them.

Coal

A number of medium-Btu coal gasification projects are already in operation. One such facility is situated in Daggett, California, where it is used for integrated gasification combined-cycle electricity production.[1] This project uses a process developed by Texaco. The Eastman plant at Kingsport, Tennessee uses the Texaco process to gasify coal for producing acetic anhydride. Other medium-Btu coal gasification facilities have been developed by Shell, Dow, and Ruhr Kohle (West Germany).[1,2]

The world's largest coal gasification facility, the Great Plains Gasification Project, was completed in 1986 at a cost of $2.1 billion.[3] This plant produces 125 million cubic feet of gas a day, which is being sold to several utilities under long-term contracts.[4] The construction and operation of this facility has yielded important know-how, which will be useful in connection with future projects.

Under present circumstances, the relatively low price of conventional gas makes such pipeline quality gasification facilities uneconomical. However, these conditions may change with the passage of time. The Department of Energy, which took over the facility in 1986, has sold it to the Basin Electric Power Cooperative of North Dakota, which is continuing its operations.[5]

Peat

Peat covers some 52 million acres in the U.S. This acreage

contains an estimated 120 billion tons of peat, which could be converted into 1,440 Tcf of methane gas. About half the peat is located in Alaska. Other large deposits are found in Minnesota, Michigan, Wisconsin, Maine, New York, and Massachusetts.[6] It is noteworthy that these states have few conventional hydrocarbon resources. Peat gasification could be a boon for these areas.

Much of the pioneering work on peat gasification has been done by Minnegasco, the gas utility serving Minneapolis, together with the Institute of Gas Technology.[7] The economics of peat gasification look promising. It has been found that peat is easier to gasify than coal.[8] Moreover, the presence of water in peat facilitates the gasification process.[9]

While considerable research has been done on peat gasification, no commercial facility for utilizing this resource has yet been constructed.

Oil Shale

Oil shale is a sedimentary rock containing kerogen. When heated, kerogen yields liquid crude oil, gases and carbon. These substances can be utilized to produce methane.

The U.S. has enormous oil shale resources. Scientists of the U.S. Geological Survey have estimated such resources at 1.3 trillion barrels of kerogen, a complex organic compound which can be converted into liquid or gaseous petroleum.[10] The Green River formation of Colorado, Utah, and Wyoming contains the largest known oil shale deposits. Other substantial shale resources are situated in the eastern U.S.

While considerable research has been done on the conversion of oil shale into usable energy, no commercial operations are currently underway. Capital costs involved in such facilities are very high. Oil shale may be considered a valuable potential resource which can supply vast amounts of oil and methane gas when conventional forms of energy have been depleted and prices have risen to higher levels.

Gasification of coal, peat, and oil shale is likely to play an increasingly important role in supplying methane gas in the twenty-first century and beyond.

Sources:

[1]The Gas Energy Supply Outlook 1987-2010, *American Gas Association, October 1987.*

[2]*"Can the Combined-Cycle Be Relied Upon to Satisfy Electricity Shortfalls on a Long Term Basis?" Douglas M. Todd,* Efficient Electricity Generation with Natural Gas, *American Gas Association and Edison Electric Institute, November 1987.*

[3]New York Times, *May 22, 1988.*

[4]*"The Great Plains Story,"* Venture, *Summer 1980, published by American Natural Resources System.*

[5]Wall Street Journal, *August 8, 1988.*

[6]*"Peat: A Major Energy Resource to Meet U.S. Clean Fuel Needs," Institute of Gas Technology, August 1979.*

[7]New Technologies for Gas Energy Supply and Efficient Use, *American Gas Association, April 1979.*

[8]*"Synthetic Natural Gas from Peat," by Arnold M. Rader,* Gas Supply Review, *American Gas Association, December 1977.*

[9]*"Peat and the Environment," Institute of Gas Technology, January 1979.*

[10]The Gas Energy Supply Outlook: 1983-2000, *American Gas Association, October 1983.*

7. Methane, a Renewable Resource

Almost all of the energy on earth comes from the sun. Much of the sun's energy is utilized by plants to convert inorganic substances into organic ones through photosynthesis. In the process, energy is stored in the plants, primarily in the form of carbohydrates (compounds of carbon, hydrogen, and oxygen). When plants have completed their life cycle, the residue is transformed into simple substances. Microorganisms play a key role in the recycling. If the conversion of plant materials takes place in an environment that excludes air, anaerobic microorganisms ingest the residue and release methane, carbon dioxide, and some other substances. The methane resulting form this process contains the energy that was originally derived from the sun.

This natural process has profound ecological significance. If plant materials were not broken down, the earth would soon suffocate in organic debris. Under those conditions, higher forms of life would become impossible. It is also noteworthy that methane is one of nature's most commonly used compounds for storing the sun's energy. This reality is the basis for considering methane a renewable resource, which will be available as long as the sun shines and plants grow on earth.

The characteristics and energy content of methane are the same, no matter what its origin. CH_4 derived from plant decomposition is the same as CH_4 found in deep sediments or produced in old oil fields and in coal seams. The energy content of pure methane is always approximately 1,000 British thermal units (Btu) per cubic foot. The lower Btu content of the gas from plant conversion is due to the fact that the methane is mixed with carbon dioxide usually in the ratio of 55 percent methane and 45 percent carbon dioxide. To obtain pipeline quality methane, the carbon dioxide is removed.

While all plant materials can be used by anaerobic microorganisms for the production of methane, the most practical sources are substances which contain large quantities of energy that would normally be wasted. These materials include garbage, manure, sewage, and ocean plants. In addition to yielding methane, the utilization of these substances brings ecological benefits.

Garbage in Landfills

In landfills, garbage is covered with soil, which keeps the air out and creates a favorable environment for the functioning of anaerobic microorganisms. The organic parts of the garbage are ingested by the microbes. Methane is one of the end products of this process.

The commercial utilization of methane gas from landfills was started in Palo Verdes, California, in 1975.[1] Since that time, many other communities have implemented methane recovery programs. Currently, about four billion cubic feet of pipeline quality methane from landfills are used by gas utilities each year.[2] In addition, methane that has not been upgraded by removing the carbon dioxide is used in several locations for electricity or power production in nearby facilities.

The utilization of methane from landfills has the following economic and environmental advantages:

(1) The landfill is usually situated within the service territory of the local gas utility. Methane from this source improves the security of the gas supply.

(2) While methane from landfills is generally more expensive than conventional natural gas, it has the advantage of avoiding transportation costs. If the landfill is properly designed, it can be a good source of methane at reasonable cost.

(3) Methane recovery from landfills usually generates taxes and royalties to the local community, as well as creating jobs and business activity. In effect, this procedure yields financial benefits from garbage disposal, which is normally a losing operation for the community.

(4) If the methane accumulated in landfills is not utilized or properly controlled, it constitutes an environmental hazard. Such methane may migrate underground and come up in inhabited areas, where it may cause fires.

(5) Insofar as methane is derived from landfills, it conserves conventional gas resources, which are non-renewable.

These advantages justify increased interest in utilizing methane from landfills wherever it can be economically recovered. Companies active in this field include GSF Energy, Wehran Energy, O'Brien Energy Systems, Waste Management, and Biogas Technology.[2] Methane Development Corporation, a subsidiary of the Brooklyn Union Gas Company, is co-owner of a methane recovery facility at New York's Fresh Kills landfill on Staten Island, one of the world's largest landfills.

Manure

Concentrated amounts of cattle manure, such as those found in large feedlots, provide a good sources of methane. Thirty-five million tons of manure are available from such sources annually. It is estimated that this manure could be converted into 200-300 billion cubic feet of methane.

The Washington Energy Company of Seattle has developed advanced technology for converting manure into methane. It utilizes special processing procedures and a proprietary fermentation process to speed up the conversion process. As a result, the company claims that its process is three times as efficient as conventional procedures. The new technology has been licensed for use by Ecotechniek of Holland. In addition to methane, the process yields animal feedstock and fertilizer.[4]

Sewage

Sewage, which consists of about 99 percent water and 1 percent organic materials, can be used to grow water hyacinths, algae, and other aquatic plants, which in turn can be transformed into methane through biodigestion by anaerobic microorganisms. In addition to generating methane, this procedure would facilitate waste management and help the environment. The commercial utilization of this approach awaits additional research and higher energy prices.[5]

Ocean Plants

Much of the earth is covered by oceans, which contain nourishment for plants and which are exposed to vast amounts of sunlight. This environment favors the rapid growth of certain plants, which can be utilized for the production of methane. For example, giant brown kelp plants (Macrocystis pyrifera), which are native to the shore off Southern California, attain a length of 200 feet. The plant grows as much as two feet in a single day. The top portions of these plants can be harvested for conversion into methane via anaerobic microorganisms. Considerable research on the use of such ocean plants has been done by the Gas Research Institute and others.[6] It has been estimated that 55,000 square miles of ocean energy farms could produce 20 trillion cubic feet of methane annually, enough to take care of all U.S. methane requirements, on a renewable basis.[7]

The large scale utilization of methane from renewable biomass sources in the U.S. awaits the development of advanced technologies and higher energy prices. It will probably be some time in the twenty-first century before these conditions prevail.

Developing Countries Lead the Way

Millions of peasants in China, India, and other developing countries recycle plant residues and other waste materials in biodigesters to produce methane and fertilizer. The biodigesters are of simple design, inexpensive, and easy to operate. They are well suited for their purpose. They utilize materials that would otherwise be wasted and they provide energy for heat, light, and cooking that is superior to any alternative.

This methane is making a major contribution to improving the standard of living of these people. It has the additional advantage of being in harmony with ecological realities. The developing countries are demonstrating the practicality of utilizing renewable sources of methane for the benefit of mankind.[8]

Sources:

[1] *"Methane from Landfills,"* Gas Energy Review, *American Gas Association, April 1980.*

[2] *"Status of Landfill Methane Recovery Projects," Sarah Doelp,* Gas Energy Review, *A.G.A., January 1988. Ibid., January 1989.*

[3] *Methane Development Corporation, 166 Montague Street, Brooklyn, New York 11201.*

[4] *"Washington Energy Finds Way to Turn Manure to Methane, Feed, and Fertilizer,"* Wall Street Journal, *March 20, 1989. For additional information, contact the Washington Energy Company, P.O. Box 1869, Seattle WA 98111.*

[5] The Gas Energy Supply Outlook: 1980-2000, *American Gas Association, October, 1980.*

[6] *"Methane from Seaweed," by J.C. Sharer and A. Flowers,* Grid, *Gas Research Institute, January 1979.*

[7] *"Marine Biomass Energy Project," by James R. Frank and Joseph E. Leone, presented to the A.G.A. Transmission Conference, Salt Lake City, Utah, May 6, 1980.*

The following additional sources on marine biomass may also be of interest:

"A Review of the Energy from Marine Biomass Program," by Armond J. Bryce, presented to the Biomass Symposium. Institute of Gas Technology, August 14-17, 1978.

"Marine Biomass Energy Project," by Joseph E. Leone, presented to the Marine Technology Society, New Orleans, October 11, 1979.

"Studies Improve Biomass to SNG Conversion," by James R. Frank, Hydrocarbon Processing, *April 1980.*

"Marine Macroscopic Plants as Biomass Source," by Wheeler J. North, UNITAR Conference on Long-Term Energy Sources, Montreal, December 1979. Professor North of the California Institute of Technology (Corona del Mar) was in charge of investigating kelp growth and nutrition for the Marine Biomass Energy Project.

[8] The following papers were presented at the UNITAR Conference on Long-Term Energy Resources, published by Pitman Publishing Inc., Marshfield, MA 02050, 1981:

"The Development of Chinese Biogas," Chen Guang-Qian and Chen Ming.

"Biogas Production in India and Other Developing Countries," P.R. Srinivasan.

"Potential of Anaerobic Digestion in Biogas Production in Developing Countries," Sermpol Ratasuk.

8. Pipelines and Storage

In 1987, U.S. gas fields were connected with more than 52 million customers by 1,151,000 miles of pipelines and mains. This distribution network was backed up by storage facilities that could hold an estimated 7.8 Tcf of gas.[1]

Gas gathering lines in the fields totaled 95,500 miles, transmission pipelines 271,900 miles, and distribution mains 783,800 miles.[1] Between 1980 and 1987, distribution mains grew from 701,800 to 783,800 miles.[1] This growth in gas mains increases the ability of the local distribution companies to reach new customers.

This network of gas gathering lines, transmission pipelines, and distribution mains provide gas with very efficient transportation. The energy required for transmission is minimal, as gas is propelled forward by compressors.

Gas usage from residential and commercial customers for space heating creates high seasonal demand in the winter months. Efficient gas production and transmission require constant operations on a year-round basis. To balance these discrepancies between demand and supply, the industry has established large storage facilities, which can hold an estimated 7.8 Tcf of gas.

The gas distribution network has considerable excess capacity. In 1972, when pipelines and mains totaled 943,000 miles, gas usage amounted to 22.7 Tcf.[1] Current gas consumption is about 4 Tcf lower and has an additional 200,000 miles of distribution facilities.

While the overall distribution network is ample to take care of current and anticipated requirements, some parts of the country need additional transmission lines to serve their customers. This need for additional gas supplies is particularly acute in the Northeast (New England and Mid-Atlantic states), which have hitherto lagged far behind the rest of the nation in gas usage.

The Northeast has a number of special characteristics, which are relevant to an understanding of its energy needs. It is a region which has few local energy sources. It is situated at the end of the pipelines that fan out from the major gas fields in the southwestern part of the country. As a result, gas transportation costs are the highest in the nation. To place this matter into perspective, in 1987 the price of gas in the U.S. averaged $4.32 per MMBtu. The cost in New England was $5.52 per MMBtu and in the Mid-Atlantic states $5.59. In contrast, the price of natural gas in Texas was only $2.78 per MMBtu.[1] The differences were primarily due to transportation costs.

Because of its location on the Atlantic ocean, the Northeast has relied heavily on imported oil for its energy requirements. The oil price escalations of the 1970's were very harmful to the region's economy. A general consensus has emerged to make greater use of alternative sources of energy. This orientation was strengthened by increased concern about environmental issues. Natural gas has emerged as the preferred choice for many residents of the Northeast. The following table shows the anticipated growth in demand for gas in the Northeast between 1987 and 1990.

Projected Natural Gas Demand Growth in the Northeast[2]
(billion cubic feet)

Customer Class	1987 Demand	1989 Demand	Increase
Residential/commercial	1,384	1,477	93
Industrial	371	663	292
Electric generation	303	761	458
Total	2,058	2,901	843

To meet this projected increase in gas demand, additional pipeline facilities are required, including more gas from Canada, which is the closest sources of supply.[3] All possible steps should be taken by federal, state, and local officials to expedite the process of supplying the gas needed by the Northeast. Such action is desirable for many reasons, including the economic vitality of the Northeast, improved environment, and strengthened energy security.

As a general principle, demand-induced gas requirements should be given priority for the construction of additional pipeline facilities. This approach makes good economic sense. Moreover, long-term planning is essential to meet future energy needs. For example, the U.S. should make plans now to bring the enormous gas resources of Alaska to the lower 48 states.

A realistic national energy policy would also make better use of the large gas storage facilities. The 7.8 Tcf capacity is equivalent to 1.3 billion barrels of oil, which is more than the Strategic Petroleum Reserve of the federal government. Natural gas can take care of most stationary energy requirements and can also play a role in reducing vulnerability to oil disruptions in the transportation sector. Mr. Eugene Luntey, former president of the Brooklyn Union Gas Company, called this eminently sensible idea to my attention.

The 1,151,000 mile gas distribution system and the 7.8 Tcf gas storage facilities are among the greatest assets the U.S. possesses in relation to economic vitality, environmental cleanliness, and energy security.

Sources:

[1] 1988 Gas Facts, *American Gas Association, 1988.*

[2] *"The Environmental and Energy Security Effects of New Pipeline Capacity to the Northeast," American Gas Association, May 16, 1988.*

[3] *"Status of Canadian/U.S. Natural Gas Arrangements,"* Gas Energy Review, *American Gas Association, May 1989.*

9. Gas Imports from Canada

Canada is richly endowed with natural gas. These resources are so plentiful that they can take care of Canadian requirements and still leave a surplus for exports. While the U.S. also has a strong natural gas resource base, it can make good use of gas supplies from Canada. Geographically, the following sections of the U.S. are in an advantageous position to utilize Canadian gas: Pacific Northwest, California, upper Midwest, Northeast and Mid-Atlantic. Canada and the U.S. have a mutual interest in increasing their natural gas trade.

Conventional reserves of Canadian natural gas have been estimated by the Canadian Petroleum Association at 94.8 trillion cubic feet (Tcf).[1] Most of this gas is situated in the provinces of Alberta, British Columbia, and Saskatchewan. In addition, an estimated 24.8 Tcf are located in frontier areas, including 14 Tcf in the Arctic Delta, 10 Tcf in the Mackenzie Delta, and 0.4 Tcf in the East Coast Offshore area.[1] A much higher estimate of potential East Coast offshore gas resources has been made by the Canadian Oil and Gas Lands Administration, which uses 10 Tcf as the estimate in its 1985 annual report.

The U.S. has imported Canadian natural gas for a long time. In the early 1970's, imports averaged one trillion cubic feet annually, or about five percent of total U.S. consumption. Imports declined to less than 800 billion cubic feet during the 1980's, but have recovered since 1987. In fact, imports reached a record 1,265 Bcf in 1988.[2] As new pipelines are constructed, it is estimated that imports of Canadian natural gas will increase significantly. They may reach 2 Tcf by the year 2000, when they would account for almost ten percent of total U.S. consumption.[1]

In 1988, imports of Canadian gas were consumed in the following areas: California, 502 Bcf (39.6%); Central (Midwestern region), 450 Bcf (35.5%); Pacific Northwest, 176 Bcf (13.9%); and Northeast, 131 Bcf (10.3%).[2] While the Northeast has lagged behind the rest of the northern tier states, it is rapidly catching up. In fact, if planned pipeline projects are implemented, the Northeast may emerge as a major growth area for Canadian gas. This prospect is enhanced by the fact that New England and the Mid-Atlantic states have hitherto underutilized natural gas, partly because of high transportation costs. (This region is situated at the end of the pipeline network from the main producing areas in the southwestern U.S.). Reasonably priced Canadian gas should enable this region to catch up with the rest of the country in making optimum use of natural gas.

It is likely that during the next several years pipeline construction to bring additional natural gas from Canada to the U.S. will reach a high level of activity. The following projects are reasonable prospects:

(1) An extension of the souther portion of the Alaska Natural Gas Transportation System (ANGTS) from lower Alberta to the Mackenzie Delta and the Arctic Delta.[3,4] Such pipeline construction would also advance the prospects for reaching Alaskan natural gas resources.

(2) Several proposals are pending to build additional capacity for delivering Canadian gas to the Northeast. If these projects are approved and implemented, Canadian gas imports to this region could more than double during the 1990's.

(3) While no specific plans have as yet been announced, it is possible that some time in the 1990's construction will start on a pipeline to bring gas from Canada's East Coast Offshore areas to the U.S. market. Such a development would have great significance for the Northeast and Mid-Atlantic states, which are not far from these gas resources. Major energy

consuming states like New York, Massachusetts, and New Jersey, as well as their neighbors, would then be *closest* to the energy sources, rather than being at the end of the pipeline system.

Overall, the future for imports of Canadian natural gas looks very bright.

Sources:

[1] *"Natural Gas Reserves of Selected Fields in the United States and Canada," prepared by the American Gas Association and the Canadian Petroleum Association as of December 31, 1988, published in June 1989.*

[2] *"Status of Canadian/U.S. Natural Gas Arrangements,"* Gas Energy Review, *Gas Supply Committee, May 1989.*

[3] *"Foothills Announces Mackenzie Valley Pipeline," Foothills Pipe Lines Ltd., Calgary, Alberta T2P 3W8, March 16, 1989.*

[4] *"Natural Gas Pipeline from Canadian Arctic to U.S. is Proposed,* Wall Street Journal, *March 20, 1989.*

10. A Long-Term Perspective on Gas Supply

The long-term outlook for natural gas supply is favorable. The large methane gas resource base in the U.S., combined with gas imports from Canada, will take care of anticipated gas demand. In fact, these resources are adequate to take care of a substantially larger market than is currently projected on the basis of conservative estimates.

In the period 1980-1987, natural gas reserve additions in the continental U.S. averaged 89 percent of production.[1] For 1988, reserve additions are estimated at 90-105 percent of production.[1] This high rate of reserve addition was achieved in spite of the fact that drilling activity was at a subdued level of fewer than 7,500 wells drilled. (This total included both oil and gas wells.) The following factors accounted for the positive performance in 1988:

(1) Drilling shifted from oil-prone formations to gas-prone ones. This fact indicated that producing companies, which are in both the oil and gas business, have reached the conclusion that drilling for gas in the U.S. makes better economic sense than drilling for oil.

This trend is likely to continue. In fact, by 1990 gas production in the continental U.S. is likely to be at least fifty percent greater than oil production.[2] The U.S. gas resource base which can be economically recovered is much larger than the oil resource base. In recent years, three times as much energy was found per gas well than per oil well.[3] These statistics send a strong message in favor of gas.

(2) The quality of drilling for gas has improved. Considerable progress has been made in recent years in geological know-how, particularly in relation to wells in deep locations (below 15,000 feet) as well as offshore. Technological developments have also improved results. Remarkable achievements were recorded in offshore drilling, with ever deeper levels of water becoming accessible to gas drilling.

Additional improvements in these areas seem likely. As demand for gas increases and drilling activity moves to higher levels, geological know-how and technologies will also benefit. Drilling activity itself, particularly in frontier areas like deep gas and offshore gas, is likely to lead to new insights and technical improvements, which, in turn, will result in more reserve additions. In the final analysis, drilling is the most reliable test for determining the true nature and scope of the resource base in any given location.

(3) Methane in coalbeds became an important area of drilling activity and resource addition in 1987-1988. The upsurge in this unconventional resource can be largely explained by the fact that the federal government has implemented a tax incentive of 70 cents per Mcf on methane produced from coal seams by January 1, 2001.[4] Increased production from this resource is very likely for the balance of this century. The impact of tax benefits on resource recovery shows that considerable quantities of unconventional gas can be produced as gas prices increase.

The following table summarizes the gas supply outlook for the period 1988 through 2010. The data are derived from the "business as usual" scenario presented in the Total Energy Resource Analysis (TERA) prepared by the American Gas Association in the spring of 1989.[5]

Gas Supply Outlook, 1988-2010
(Quadrillion Btu)

Year	Domestic Production	Imports	Alaska & Misc.	Total
1988	16.7	1.3	0.5	18.6
1989	17.4	1.2	0.5	19.1
1990	17.4	1.3	0.5	19.2
1995	18.6	1.6	0.5	20.7
2000	17.8	2.8	1.0	21.6
2005	17.2	3.1	1.9	22.2
2010	17.1	3.5	1.9	22.5

These projections indicate that the bulk of gas supplies throughout the period 1988-2010 will be produced in the continental U.S. Imports, primarily from Canada, will more than double. Alaskan gas will make a significant contribution some times after the year 2000.

The data in this table are based on conservative assumptions about gas prices. If prices increased substantially, gas production in the U.S. could be stepped up by 5 Tcf or more during this period. A large part of the increased supply would come from such sources as tight sands, Devonian shale, geopressured brine, coal seams, and biomass. Technological developments in connection with some of these and other unconventional gas resources may make it possible to achieve production increases without a major run-up in prices.

It is noteworthy that the high-price scenario, which would most probably result from substantial increases in world oil prices, would improve the prospects for U.S. energy independence via the gas option. Five Tcf of methane gas would replace approximately 2.5 million barrels of imported oil a day. If the U.S. implements realistic energy policies, the gas option could play a major role in minimizing adverse economic effects from international oil price escalations.

Sources:

[1] *"Preliminary Findings Concerning 1988 Natural Gas Reserves,"* Energy Analysis, *American Gas Association, April 28, 1989.*

[2] *"Global Climate Change and Emerging Technologies for Electric Utilities: The Role of Natural Gas," Nelson E. Hay, American Gas Association, 1988.*

[3] The Power of Natural Gas, *Catalog No. G92237, American Gas Association, 1988.*

[4] The Gas Energy Supply Outlook 1987-2010, *American Gas Association, October 1987.*

[5] *"The 1989 A.G.A. Total Energy Resource Analysis (TERA) Base Case,"* TERA Analysis, *American Gas Association, April 7, 1989.*

PART II
GAS MARKETS AND TECHNOLOGIES

11. Expanding Markets for Gas

In the period 1987 to 2010, it is estimated that demand for natural gas will increase from 17.7 to 23.1 quadrillion Btu (Quads), a rise of about thirty percent.[1] The following table provides details on estimated major markets for natural gas.

Natural Gas Markets, 1987-2010[1]
Quadrillion Btu

Year	Residential	Commercial	Industrial	Electric Utilities	Misc.	Total
1987	4.4	2.5	7.3	2.9	0.5	**17.7**
1988	4.8	2.7	7.9	2.7	0.4	**18.5**
1989	4.7	2.9	8.1	2.8	0.5	**19.0**
1990	4.7	2.9	8.3	3.1	0.5	**19.5**
1995	4.9	3.3	8.9	3.7	0.5	**21.3**
2000	5.0	3.6	9.2	3.9	0.6	**22.2**
2005	5.1	3.8	9.1	4.2	0.5	**22.6**
2010	5.1	3.9	9.0	4.5	0.5	**23.1**

In 1987, the gas utility industry had 47.4 million residential customers. They used 4.4 Quads of natural gas. Space heating accounted for 66 percent of the residential gas load. The other main residential gas uses included water heating (24%), cooking (5%), and clothes drying (2%).[2]

Overall, residential gas consumption is expected to rise from 4.4 Quads in 1987 to 5.1 Quads in 2010, an increase of 16 percent, which averages out to less than 1 percent a year. The main cause for this slow growth is the upsurge in efficiency in gas appliances and energy-conserving building construction. Since 1973, such efficiency improvements have averaged around 1.7 percent annually.[3] This conservation trend is expected to continue throughout the period 1987-2010. In fact, it may be accelerated by the requirements of the National Appliance Energy Conservation Act, which mandates high efficiency ratings for new gas burning furnaces and water heaters in 1992 and subsequent years. It is noteworthy that if it were not for conservation and improved efficiency, residential gas consumption by the year 2010 might be about 7 Quads, or 40% higher than the figure projected in the preceding table.

While the overall residential market is estimated to remain essentially flat during the next two decades, some market sectors are expected to show significant growth. Natural gas is likely to penetrate into the cooling market with air conditioning units and gas heat pumps. This market has the advantage of increasing the gas load during summer months, when most of the gas distribution network is underutilized. Cost advantages have enabled gas to increase market share in the water heating market, a trend that is likely to continue. Both of these developments will improve the overall performance of the gas industry, which needs to balance its load on a year-round basis for more cost-effective operations.

In 1987, the U.S. gas utility industry served 4.0 million commercial customers. This market includes office buildings, retail trade, health services, educational facilities, and restaurants. They consumed a total of 2.5 quadrillion Btu of natural gas. Space heating accounted for 62% of the total. Other uses included water heating (15%), cooking (14%), process heat (2%), cooling (1%), cogeneration (1%), and miscellaneous (5%).[2]

Rapid growth is projected for the commercial gas market, which is estimated to increase from 2.5 quadrillion Btu in 1987 to 3.9 in 2010. The fastest growth is expected in gas cooling and cogeneration, both of which will improve the year-round load utilization of the natural gas industry.

In 1987, gas consumption for industrial applications totaled

7.3 Quads. A large part of this gas was bought directly by end users from non-utility sources. Gas utilities supplied 2.3 Quads of the total. [2] According to the estimates in the table at the beginning of this chapter, gas consumption in the industrial market is likely to increase to 9.0 Quads by 2010. The industrial market for methane gas may be subdivided as follows:

Industrial Gas Markets, 1987-2010[1]
(Quadrillion Btu)

Year	Manufacturing	Lease & Pipeline Fuel	Cogenerators & IPP's*	Total
1987	5.7	1.2	0.4	7.3
1988	6.2	1.2	0.5	7.9
1989	6.0	1.2	1.0	8.1
1990	6.0	1.2	1.1	8.3
1995	5.6	1.2	2.2	8.9
2000	5.0	1.1	3.1	9.2
2005	4.3	1.1	3.6	9.1
2010	3.7	1.1	4.2	9.0

*IPP's refers to Independent Power Producers

(Note: Some of the totals may be slightly different from the addition of the categories because of rounding out for statistical purposes.)

The manufacturing market represents conventional boiler and process applications. The projections anticipate a decline in this market after 1990. This decline is more than offset by the large increase in cogenerators and independent power producers, whose demand for natural gas is expected to increase about tenfold between 1987 and 2010. These applications produce electric power and process heat.

The use of gas by electric utility power plants is expected to grow from 2.9 Quads in 1987 to 4.5 Quads in 2020. New technologies, notably combined cycle, which improves efficiency by making use of waste heat for additional electricity production, are expected to capture larger markets to meet growing demand for electricity. In addition, the select use of gas with coal will increase sharply to help utilities meet environmental standards.

The data used in this chapter are based on conservative, business-as-usual assumptions. While the anticipated growth in overall gas consumption is relatively modest, it is quite impressive if one considers the strong trend toward conservation and improved energy efficiency which will probably characterize this era. To place this matter into perspective, if one assumes annual energy efficiency improvements of 1.7% throughout this period, gas consumption will be held back as much as 10 Quads between 1987 and 2010. While improved energy efficiency and conservation slow down the growth of natural gas (and all other forms of energy), they are in the best interest of consumers and of the nation. The natural gas industry has recognized this reality all along and has actively promoted the development and marketing of high efficiency gas appliances.

During the next two decades, the gas industry will find growing applications of natural gas for cooling, cogeneration, powering vehicles, and combined cycle electricity production. These technologies will help create a more balanced load for gas suppliers throughout the year, rather than being excessively dependent on the heating season. As a result, the gas industry will make more efficient use of its capital-intensive facilities, which should bring improvements in operations and profitability.

Sources:

[1] *"The A.G.A. TERA 1989 Mid-Year Base Case," American Gas Association, August 16, 1989 (draft).*

[2] Gas Facts: 1987 Data, *American Gas Association, 1988.*

[3] *"Peak-Quarter Conservation in the Residential Market: 1973-1986,"* Energy Analysis, *American Gas Association, September 18, 1987.*

12. Gas Heats Best

Natural gas provides comfortable heat from efficient equipment at reasonable cost. A wide range of gas furnaces and boilers is available from many manufacturers. Recent technological developments have enabled gas heating equipment to reach efficiencies of 90 percent or more. Most homeowners can achieve considerable savings if they switch to efficient gas heat from competing energy sources.

At the end of 1987, 55 percent of all existing homes were heated with gas, compared with 21 percent electric, 14 percent oil, and 10 percent others.[1] The popularity of gas for heating is evident from the fact that when gas mains reach areas served by competing energy sources, many homeowners switch to gas.

The performance of heating equipment is judged on the basis of its Annual Fuel Utilization Efficiency (AFUE). This yardstick measures how much useful heat is supplied by the amount of fuel consumed. For example, an AFUE rating of 50 means that half the fuel is transformed into useful heat, while the other half is wasted.

In recent years, gas heating technology has allowed the AFUE to increase from around 60 to 90 and above. This remarkable improvement in efficiency involved the following changes:

(1) Conserving heat in the equipment through insulation and the use of suitable materials and construction techniques.

(2) Replacing the pilot light with electronic ignition.

(3) Developing highly efficient combustion technologies.

(4) Utilizing very effective heat exchangers.

(5) Extracting heat from flue gases, using it for space heating rather than having it go up the chimney.

As a result of these technological advances, 90 percent or more of the heat energy contained in each cubic foot of gas is available to heat the dwelling. Moreover, this type of equipment does not require a chimney. Venting of the flue gases can be done through a plastic pipe via a sidewall.

High-efficient gas furnaces and boilers, in a wide range of sizes and models, are available from 37 U.S. manufacturers. Two recent articles are recommended for getting a good overview of what equipment is currently available: (1) "Your Guide to Selecting a New Furnace," by Jeffrey Seisler (*Consumers Digest,* September-October 1988); and (2) "1988-89 Guide to High-Efficiency Furnaces," by Michael S. Weil (*Contracting Business,* August 1988).

The Seisler article provides a general introduction to high-efficiency gas furnaces and presents detailed information about a dozen types. Prices of these furnaces range from $1,300 to $3,000. This article also offers useful tips in connection with the purchase of a furnace, including the suggestion that the prospective purchaser should get at least three estimates from different contractors.

Mr. Weil's article has illustrations of some 24 furnaces, along with information about their manufacturers, description of their capabilities, efficiency ratings, technical data, average installed costs, and how they work.

Most homeowners who are connected to gas mains would be well advised to consider switching from old oil-burning equipment or from most electric heat installations to high-efficiency gas furnaces or boilers.

Additional information may be obtained from the *1989-90 Guide to High Efficiency Furnaces,* published by *Contracting Business.* Another useful source is "The Right Choice—Energy Efficient Gas Heating Systems," by the American Gas Association (Catalog No. B50139).

When oil-burning equipment is more than twenty years old, it is likely that its AFUE rating is less than 50. By switching to high-efficient gas equipment, substantial dollar savings can be realized. Moreover, old oil furnaces and boilers tend to require frequent maintenance and repairs, which add to the cost. Independent contractors and the local gas utility can provide detailed information on estimated savings.

Electric heat costs substantially more than gas heat in most parts of the country. A study by the American Gas Association estimates that homeowners can save $400-$1,000 a year by switching from electric heat to a high-efficiency gas furnace.[2] Assuming such a furnace to cost $2,000, the annual return on investment would range from 20-50 percent. Over a twenty year period, which is the normal life span of a furnace, these savings could total from $8,000 to $20,000.

It also makes good sense to switch from an old gas furnace to a new high-efficient one. Twenty-year-old gas equipment usually has an AFUE rating of 60 or less. The new high-efficient furnace will save one-third or more on fuel bills.

In 1987, residential customers used 4.4 Quads of natural gas.[3] Sixty-six percent of this total or 2.836 Quads, were used for space heating.[1] In the same year, commercial customers' gas use totaled 2.4 Quads,[3] of which 62 percent, or 1.438 Quads were devoted to space heating.[1] Combining the two sets of numbers, it is apparent that 4.436 Quads of gas was used for space heating. Generally speaking, the heating season lasts about six months each year. The gas utilities have to maintain facilities to supply these requirements on a year-round basis. Considering the capital-intensive nature of the gas business, it makes good economic sense for the utilities to encourage customers to switch to high-efficient heating equipment, which helps to reduce the unbalanced load.

The trend to smaller homes and tighter, more insulated building construction, makes possible the use of heat sources smaller than a full-sized furnace. A number of manufacturers have developed space heating systems that use the hot water heater to deliver heat. This procedure can also be adapted to air conditioning. These devices may be considered as replacements for electric heat pumps. For additional information about such appliances, check with the contractor and/or with the gas utility about integrated water heating/space heating systems.

Natural gas has been the best source of space heat for more than a century. From all indications it is likely to retain that status for a long time to come. In fact, the development of high-efficiency gas heating equipment, together with the comparatively advantageous cost of gas, make it likely that gas will gain greater market share in the foreseeable future.

Sources:

[1] Gas Industry Five-Year Strategic Marketing Plan, 1989-1993, *American Gas Association, December 1988.*

[2] *"Cost Advantages of Natural Gas in Existing Homes,"* Energy Analysis, *American Gas Association, April 4, 1986.*

[3] *"The 1989 A.G.A. TERA Base Case,"* Total Energy Resource Analysis Model, TERA-Analysis, *American Gas Association, April 7, 1989.*

13. Heating Water with Gas Achieves Big Savings

The production of hot water requires substantial amounts of energy on a year-round basis. Natural gas is the most cost-effective fuel for this purpose. Consumers who currently use electricity or fuel oil to produce hot water could save billions of dollars annually by switching to high-efficiency gas appliances.

In 1987, hot water was produced in 86.3 million households, with the following breakdown by fuel source (millions of units):[1]

Natural gas, 46.9

Electricity, 28.9

Fuel oil and kerosene, 5.4

Other, 5.1

Fifty-four percent of all households use natural gas for hot water production. The highest percentages of gas water heaters are found in the Midwest and the West. In contrast, the South relies heavily on electric water heaters. Most of the fuel oil and kerosene units are situated in the Northeast.

The production of electricity involves very high capital expenses. Moreover, the conversion of fossil fuels (coal, natural gas, or oil) into electricity causes the loss of fifty to seventy percent of the energy contained in the original source. Consumers of electricity have to pay for all these costs. In contrast, capital costs related to natural gas delivered to households are modest and all the energy contained in the gas is available for end use application. As a result, the cost of energy derived from electricity is substantially higher than that of natural gas. According to data compiled by the Gas Appliance Manufacturers Association,[2] the conversion from a conventional electric water heater to a high-efficiency gas system can save more than $300 in operating costs annually. Even the switch from a high-efficiency electric system to its natural gas counterpart can bring annual operating savings of more than $200 in many instances.[2] New gas water heaters installed in the home can be purchased for as little as $500. It is apparent from these data that the return on this investment is likely to be very high. Because the costs of electricity and natural gas vary in different parts of the country, consumers should check the facts applicable to their situation with appliance dealers and/or their local gas utility.

The use of fuel oil for hot water production is largely confined to the Northeast (New England and Middle Atlantic states). Prior to 1973, fuel oil was the cheapest energy source available to consumers in this region. As a result, they relied heavily on this fuel for their energy needs, including hot water production. In fact, fuel oil was so cheap that in most cases they used the same oil boiler for heating the home and for producing hot water. This procedure wastes a great deal of energy, particularly in the summer months, when the large boiler is used solely for producing hot water. The upsurge of fuel oil prices since 1973 has proved to be very costly to households relying on this obsolete approach for producing hot water. They would be well advised to switch to a gas water heater to meet their requirements for hot water. In view of the fact that most of the oil boilers are more than twenty years old, and have relatively low efficiency levels, such consumers can save even more money by switching to an efficient gas unit for heating their homes.

The option to switch from electricity or fuel oil to natural gas for hot water production is particularly attractive to those households which are already connected to gas mains. Some eight million households fall in this category, including 3.2 million in the Northeast and 2.6 million in the South.[3] In those instances, the gas hot water appliance can be installed as soon as the homeowner decides to make the conversion. Assuming the annual savings per household to range form $200-$300, these eight million customers could in the aggregate save $1.6-$2.4 billion each year. These savings can be realized on an investment in gas water heaters estimated at $500 per unit, or $4 billion for the eight million households involved. On that basis, the return on investment ranges from forty to sixty percent annually over the life of the appliance, which is generally in the ten to fifteen year range. It should be emphasized that these data are broad averages for the nation as a whole. To obtain reliable information applicable to a specific situation, each consumer contemplating a switch should check with appropriate sources, including appliance dealers and local gas utilities.

The twenty-six million households using electricity or fuel oil for hot water production, which are currently not connected to gas mains have to consider the feasibility and costs involved in making such connections to gas distribution lines. These matters should be discussed with the local gas utility.

Appliance manufacturers will be able to meet the demand for gas water heaters. Shipments of such appliances have increased steadily from 2.5 million units in 1981 to 3.7 million units in 1987.[4] Apparently, an increasing number of homeowners have become aware of the advantages of heating water with gas.

Sources:

[1] Residential Energy Consumption Survey, *Energy Information Administration, May 1987.*

[2] Consumers' Directory of Certified Water Heater Efficiency Ratings, *Gas Appliance Manufacturers Association, July 1987.*

[3] *"New Demand Opportunities in Residential Gas Water Heating,"* Energy Analysis, *American Gas Association, November 13, 1987.*

[4] 1988 Gas Facts, *American Gas Association, 1988.*

14. Gas Cooling Benefits Consumers and Industry

More than sixty percent of natural gas usage in residential and commercial markets is devoted to space heating, while cooling accounts for less than 1 percent of the total.[1] In 1987, the space heating load of the gas industry amounted to 4.325 *trillion* cubic feet (Tcf).[2] In contrast, sales of gas for cooling were less than 180 *billion* cubic feet (Bcf). These data show graphically that the natural gas load varies widely with the season. In cold weather, gas distribution facilities are usually fully employed, while in warm weather they are grossly underutilized.

It makes good economic sense for the gas industry to place greater emphasis on penetrating the cooling market. The better load balance that would follow in the wake of such action would improve the overall efficiency and profitability of the gas industry.

Leaders of the gas industry are aware of these realities. In June 1989, Ralbern H. Murray, Vice Chairman of Consolidated Natural Gas Company, proposed the following targets for gas cooling: (1) In the residential market, gas heat pumps should be installed in 4.8 million homes over the next ten years; (2) Commercial gas cooling equipment should reach a level of 500,000 tons a year, or ten percent of the total market, by the year 1995.[3]

To place these numbers into perspective, in 1987 only 107,000 residential gas cooling units existed.[4] The commercial gas cooling market totaled 977,000 tons.

The following companies currently manufacture gas cooling equipment:[5]

Gas-Fired Absorption Cooling

Company	Equipment	Size
American Yazaki Corporation Dallas, Texas	double-effect chiller/heater	7.5-100 tons
Dometic Corporation Evansville, Indiana	single-effect chiller	3-5 tons
Hitachi, c/o Gas Energy Inc. Brooklyn, New York	double-effect chiller/heater	100-1500 tons
Sanyo/Bohn Heat Transfer Danville, Illinois	double-effect chiller/heater	20-950 tons

Natural Gas-Engine Driven Cooling

Company	Size
American Utility Control Tempe, Arizona	75-350 tons
Hercules-Energy Products Div. Canton, Ohio	65-135 tons
Gemini Engine Company Corpus Christi, Texas	30-150 tons
Econochill, Inc. Shingle Springs, California	50-125 tons
TecoChill, c/o Tecogen, Inc. Waltham, Massachusetts	150-450 tons
Thermo King Corporation Minneapolis, Minnesota	15 tons

Natural Gas Dessicant Systems

CargoCaire Engineering Corp. Amesbury, Massachusetts		15-80 tons
Kathabar Systems Division Somerset Technologies, Inc. New Brunswick, New Jersey		5-500 tons
A.S.K. Waco, Texas	combination/dessicant cogeneration system	2-6 tons; 35 KW, 240 MBtu
International Cogeneration Corp. Philadelphia, Pennsylvania	combination/dessicant cogeneration system	up to 60 tons; 150 KW, 540-940 MBtu

The preceding information on gas cooling equipment manufacturers was supplied by the Marketing Department of the American Gas Association, which maintains an up-to-date file on this subject. The Gas Research Institute is another source of information on developments in relation to gas cooling equipment, particularly those involving research on new technologies.

A recent study by the American Gas Association[8] shows that commerical gas cooling equipment can offer substantial savings over its electric counterpart, particularly where electric rates are high. The study was focused on medium-sized commercial buildings which can be cooled with 150 ton gas-engine driven chillers. The following table shows the estimated equipment paybacks in various cities.

Gas Equipment Paybacks
(years)

Engine-driven 150 ton gas vs. reciprocating electric:

Atlanta	Phoenix	Dallas	Chicago	Los Angeles*
3.2	2.9	7.7	3.9	3.8
Gas unit with $100/ton rebate:				
2.2	2.0	5.4	2.7	2.7

*Assuming the current (1989) Southern California Edison GS-2 rate schedule

The savings on gas cooling versus electricity are sufficiently large to provide a very attractive payback schedule in all cities except Dallas, which has relatively low electric rates.

The inducement for using gas equipment would be increased substantially if the gas utility offered a $100/ton rebate, or $15,000 per installation of the 150 ton gas-engine driven chiller. Such a rebate would facilitate the financing of the equipment, as well as increasing the return on investment by the customer. While each utility makes its own determination on rebates, if any, such action can generally be justified in terms of balancing the gas load during the summer months. The procedure is analogous to filling empty seats on an airplane at discount prices. Rebates are particularly useful when new equipment is being introduced and when the gas industry is trying to break into the cooling market in a big way.

While the main marketing thrust in the near future is likely to be in the commercial gas cooling field, the residential market should not be overlooked. The following companies are developing gas-fired heat pumps that can provide cooling as well as heating: (1) Aisin Seiki Co., Ltd.; (2) York International Corporation; (3) Phillips Engineering Company, Inc.; and (4) Columbia Gas System Service Coporation.[7]

The Aisin gas heat pump is a natural gas engine-driven unit sized for a 5-ton cooling capacity and heating of 90,000 Btus per hour. Aisin Seiki Co. is a leader in gas heat pump technology in Japan, where it has gained one-third of the market. The company is working closely with the Gas Research Institute to meet U.S. specifications for its equipment. A field test of ten prototype heat pumps is scheduled for the fall of 1989.[7]

The York gas heat pump is being developed by the company's engineers with support from Battelle Memorial Institute, Briggs & Stratton Corporation, and the Gas Research Institute. One of its unique features is a single-cylinder engine. This engine utilizes hydraulic lifters and special wear-resistant materials for valve seats, rings, and ring grooves. The engine has been designed for a service life of at least 40,000 hours. It can develop up to five horsepower to produce a cooling capacity between 2.5 and 4 tons, and a heating capacity of 50,000 Btus per hour. A field test for this gas-fired heat pump is scheduled for the fall of 1989.[7]

Phillips Engineering is developing a gas-fired heat pump with a new absorption cycle that is considerably advanced over existing types used in residential refrigeration and air conditioning. The U.S. Department of Energy and the Gas Research Institute are playing a role in the development of this heat pump.[7]

Columbia Gas has developed a prototype of a residential double-effect gas-fired absorption heat pump that will function as a chiller/heater. The unit has a target capacity of three refrigerant tons of cooling and 70,000 Btu per hour of heating output. Columbia is seeking a manufacturer to commercialize this heat pump.[7]

The high caliber of the companies and research institutes involved in developing these gas-fired heat pumps for residential application offers good prospects that this type of equipment will soon become available.

Most people have been conditioned to think of gas primarily, if not solely, for heating purposes. Recent technological developments indicate that gas cooling has considerable potential that may eventually balance the heating functions.

Sources:

[1] Gas Industry Five-Year Strategic Marketing Plan 1989-1993, *American Gas Association, December 1988.*

[2] *"The 1989 A.G.A. TERA Base Case,"* TERA Analysis, *American Gas Association, April 7, 1989.*

[3] *Remarks by Ralbern H. Murray, Vice Chairman, Consolidated Natural Gas Company, to the Managing Committee of the Space Conditioning Commercialization Group, Hilton Head, South Carolina, June 17, 1989.*

[4] Gas Househeating Survey: 1987, *American Gas Association, July 1988.*

[5] *Names of gas cooling equipment manufacturers were supplied by the Marketing Department of the American Gas Association.*

[6] Gas Cooling, the Opportunity is Now, *American Gas Association, 1987.*

[7] *"Space Conditioning Commercialization Group Gas Equipment Display," information material prepared by the Gas Research Institute in connection with the display at Hilton Head, South Carolina, June 1989. For additional information, contact the Gas Research Institute, 3600 West Bryn Mawr Avenue, Chicago, Illinois 60631.*

[8] *"An Analysis of the Economics of Gas Engine-Driven Chillers,"* Energy Analysis, *American Gas Association, May 26, 1989.*

15. Cogeneration, a Growing Market for Gas

The conventional generation of electricity with fossil fuels involves the loss of about two-thirds of the fuel's energy content in the form of waste heat. Cogeneration systems are designed to make productive use of the waste heat, thereby improving the overall energy efficiency of the fuel conversion process. Because heat energy, in the form of steam or hot water, can be economically transported only over a limited area, cogeneration facilities are most practical on sites that can make optimum use of heat. Cogeneration facilities are usually linked to an electric utility, which makes possible the sale of excess electricity and/or the purchase of backup power when needed. While all fossil fuels can be used for cogeneration, natural gas has significant cost and environmental advantages in many applications.

A gas-fired cogeneration system is driven by a prime mover, which converts the fuel into mechanical power. This function can be fulfilled by reciprocating engines, combustion gas turbines, and steam turbines.[1] The prime mover drives a generator, which produces electricity. The waste heat that accompanies this fuel conversion is captured by a heat recovery device, which transforms the heat to usable energy in the form of steam or hot water. Larger cogeneration systems are generally custom built for a specific site. Factory packaged systems are also available; they are particularly practical for smaller applications.[2] Cogeneration systems range in size from 2.2 kilowatts to as large as several hundred megawatts.[3]

The commercial markets for gas-fired cogeneration systems include hospitals, schools and universities, hotels, apartment house complexes, and computer centers. Industrial applications are widespread among industries using large amounts of electricity and heat, such as pulp and paper, food processing, textiles, oil refining, fertilizer, petrochemicals, and pharmaceuticals.[3]

Many factors must be considered in determining the economic feasibility of a cogeneration system, including prospective electric and heat energy requirements, fuel costs, capital costs, arrangement with the electric utility for the sale of excess electricity and/or the use of backup power, operating expenses, and taxes. It is advisable to make use of expert guidance in connection with the investigation of this option. Most natural gas utilities have people on their staff who are specifically trained in cogeneration technology and economics. Equipment manufacturers may also be helpful. Independent engineering studies can be useful in analyzing alternative options on an unbiased basis; they should generally be utilized for larger installations. A properly designed cogeneration system should make good economic sense. As a rough guideline, capital costs should be recovered in a reasonable period of time, generally five years of less.

The environmental cleanliness of natural gas is particularly important in cogeneration facilities. Commercial applications generally involve sites with concentrated numbers of people, who require clean air. Similarly, many industrial uses involve processes that make a clean environment essential. Studies indicate that a gas-fired steam turbine cogeneration unit capable of providing 200,000 pounds of steam per hour emits only 6 percent of the sulfur dioxide, 10 percent of the particulate matter, and 10 percent of the nitrogen oxides of a coal burning cogenerator equipped with pollution control equipment.[3]

Because of their economic and environmental advantages, natural gas-fired cogeneration facilities are in a rapid growth phase. Existing cogeneration capacity utilizing natural gas as fuel totals 17.3 gigawatts. An additional 13.5 gigawatts are under construction or planned for the next several years. In 1987, about 800 billion cubic feet of natural gas were used for cogeneration. Estimates for the early 1990's range from 1.3 to 2 trillion cubic feet.[4]

The use of cogeneration varies greatly among regions in the U.S. Currently, the greatest concentration of cogeneration is in the West South Central region, which has a large number of industries that can take advantage of waste heat recovery. The other major market is the Pacific region, particularly California, which uses cogeneration for enhanced oil recovery and which has regulatory bodies that have encouraged cogeneration for environmental reasons. Together, these two regions account for 88 percent of gas-fired cogeneration facilities currently in place in the U.S.[4] Based on planned cogeneration additions, the greatest growth over the next several years will take place in the Mid Atlantic, East North Central, and New England regions.[4]

The strong growth in gas-fired cogeneration systems is based on sound economics, beneficial environmental impact, and realistic long-term thinking about energy availability and costs. These considerations are likely to continue favoring such systems for years to come.

Sources:

[1] *J.A. Orlando, P.E., GKCO, Inc., "Cogeneration Design and Operating Options," Chapter 9 of* Guide of Natural Gas Cogeneration, *Nelson E. Hay, Editor, The Fairmont Press, Lilburn, Georgia, 30247, 1988.*

[2] *Ibid., Chapter 10, "Factory-Assembled Systems."*

Guide to Natural Gas Cogeneration *is the most comprehensive treatment of this topic currently available. For additional information, contact The Fairmont Press, 700 Indian Trail, Lilburn, Georgia, 30247.*

[3] *"Natural Gas Cogeneration Systems,"* Gas Energy Update, *American Gas Association, October 1988.*

[4] *"Estimated Current Natural Gas Cogeneration Capacity: 1988 Update,"* Planning & Analysis Issues, *American Gas Association, March 28, 1988.*

The information cited in sources 3 and 4 was brought up-to-date as of June 1989 in a telephone conversation with Paul P. McArdle, Policy Analyst, American Gas Association. He also sent me the Energy Analysis entitled "An Economic and Environmental Comparison of Natural Gas and Coal Use for Industrial Cogeneration—1989 Update," American Gas Association, June 9, 1989.

16. The Choice Fuel for Efficient Electricity Generation

Over the next several years, electric utilities in the United States are faced with major challenges. They must meet increased demand for electricity. They must reduce environmental pollution. They must keep operating and capital costs under control. Natural gas is the fuel of choice for helping to cope with all of these challenges.

Historically, demand for electricity has grown at a rate exceeding the growth of the economy as a whole. For example, from 1965 to 1974, summer peak demand for electricity grew at an annual rate of 7.2 percent.[1] The oil price escalation of 1973 initiated a period of rising electricity costs and rates, which induced conservation. Demand slowed to an annual growth rate of three percent. During the recession of 1982, electricity sales actually declined by 2.8 percent, the first absolute reduction in electricity use since 1945. More recently, peak demand has shown substantial growth, increasing by 3.6 percent in 1986 and 4.0 percent in 1987.[1] On the basis of conservative assumptions, peak electricity demand is likely to grow by at least 1.9 percent a year for the period 1988-1997. Some estimates range as high as 2.9 percent average annual growth for this period.

Currently, 32,500 megawatts of electric generating capacity are under construction. An additional 40,000 megawatts are planned. Even if all these projects were completed in time to meet the minimum anticipated growth in electricity demand, a shortage of 10,000 megawatts would occur by 1997. If the demand for electricity grows by 2.9 percent annually, the shortfall by 1997 would be about 65,000 megawattts.[1]

The low demand scenario would require substantial electric capacity increases in Texas, the Southwest Power Pool, Louisiana, Arkansas, Oklahoma, Kansas, western Missouri, western Mississippi, California, southern Nevada, and New England. The high demand growth estimate (2.9 percent annually) would lead to shortfalls in electric generating capacity in virtually all regions of the United States.[2]

Prior to 1973, the most likely option for meeting the anticipated increase in electricity demand would have been the construction of large base load facilities, using coal or nuclear energy as fuel. Presently, this option has lost most of its appeal. Since 1980, plans for constructing coal-fired base load facilities have dropped from 100,000 megawatts to 15,400 megawatts in 1988. No new nuclear facilities have been ordered since 1978.[1]

The reasons for the declining interest by electric utilities to build new base load plants fueled by coal or nuclear energy may be summarized as follows:

(1) Their capital costs are very large, ranging from several hundred million dollars to several billion dollars, depending on capacity.

(2) It generally takes ten or more years to plan, design, and construct a coal-fired electric generating facility. Nuclear plants take even longer.

(3) The construction of such facilities locks the utility into a framework that may not be appropriate by the time the new capacity comes on stream. This lack of flexibility can be very costly.

(4) Coal has major environmental problems, which are of increasing concern. Nuclear energy has fallen into such ill repute that several facilities were closed under pressure from public authorities.

(5) Regulatory agencies have engaged in second-guessing management decisions on base load facilities which came on

stream after economic or other conditions had changed.[3]

Taking all these factors together, it is understandable that utility managements are reluctant to take the risks involved in constructing giant new facilities.

Environmental considerations also play an increasingly important role in relation to electricity generation. Coal-fired facilities face the challenge of meeting environmental standards in a cost-effective manner.

Electric utilities have several options for meeting future electricity demand growth and environmental restrictions: (1) Prolonging the useful life of existing coal-fired generating facilities, while upgrading their environmental performance; (2) Constructing combined cycle plants on a modular basis; and (3) Purchasing electricity from independent suppliers. Most utilities will probably employ all of these approaches.

Prolonging the Life of Existing Coal-Fired Facilities

The normal life span of coal-fired electric generating facilities is between 30 and 40 years. When they reach age 30, their fuel efficiency declines by 6-8 percent. Moreover, their forced outage rate increases significantly with the passage of time, while maintenance and repair costs go up. By 1990, about one-fourth of all coal-fired plants, or approximately 75,000 megawatts, will reach the critical age of 30.[4]

Combined cycle repowering is a technique for prolonging the life of old coal-fired facilities. This approach refers to the integration of new gas-fired equipment with existing coal-fired units. The basic components of such a powerplant may include a gas turbine/generator, a waste heat recovery unit, and a steam turbine/generator. The final configuration will be similar to a gas-fired combined cycle plant.

By converting waste heat into electricity, usable power per unit of fuel is increased. Repowering of old coal-fired facilities with gas equipment has the following advantages over other options: (1) Lower capital costs; (2) Shorter construction lead times; (3) Improvements to the environment; and (4) Modular design, making it possible to add capacity as needed.[4]

The specific applications of this approach will vary, depending on the nature and requirements of each facility. The following repowering procedures are technically feasible: (1) Peaking turbine repowering; (2) Heat recovery repowering; and (3) Boiler repowering.

Many coal-fired electric generating facilities already have gas turbines, which are used for meeting peak power requirements. About seven percent of installed capacity consists of such gas-fired turbines. Many of these turbines are used fewer than 100 hours a year.[7] These gas turbines can be converted to base load operation by adding a heat recovery steam generator, air-cooled condenser, and steam turbine/generator. In its repowered form, the gas turbine will become a steady source of electric power and its waste heat will be converted into additional electricity.[4]

A second procedure is to replace the old coal-fired boiler with a new gas turbine/generator and a heat recovery steam generator. This approach makes use of the existing steam turbine/generator to produce additional electricity.[4]

A third option is to repower the main coal-fired boiler by adding a gas turbine/generator. The exhaust from the turbine/generator provides combustion air to the existing coal boiler, which replaces the function of the existing forced-draft fan and air heater. The gas turbine and coal boiler also complement each other in the reduction of nitrogen oxides. This lowering of a major pollutant is achieved by reducing the flame temperature in the coal boiler, and by adjusting the fuel mix in the burners to accelerate decomposition of nitrogen oxides. Boiler repowering is widely used in Europe, but it is also gaining increased interest in the U.S.[4]

All of these repowering procedures can be more economical than the costs involved in constructing new coal-fired capacity. The environmental benefitss are also very impressive. The following table compares the emissions from heat recovery repowering with a new coal-fired facility. Data are for a 198 megawatt plant.[4]

Pollution Emissions
(tons/year)

	Heat Recovery Repower	New Coal-Fired Powerplant
Sulfur Dioxide	2	3,395
Nitrogen Oxides	1,290	1,980
Particulates	9	170
Sludge	0	138,075
Ash	0	49,320

The virtual absence of sulfur dioxide emissions from the gas-fired turbine is particularly noteworthy, for this pollutant is considered to be the major cause of acid rain. The absence of sludge and ash from the gas equipment reduces operation costs and eliminates the problem of finding dumping grounds for these materials.

Natural Gas Can Help Coal

The select use of natural gas with existing coal-fired electric generating facilities can help meet environmental standards while improving operating performance in a cost-effective manner. This approach involves minimal capital expense.

The traditional procedure utilizes a relatively small amount of gas (usually around 10 percent) which is cofired with the coal. This combination improves combustion, increases operating efficiency, and reduces pollution. Many electric utilities make use of this technique.

A more advanced technology is called "reburning." This procedure involves the injection of natural gas at a high elevation in the coal-burning facility, where the combustion gases are mixed with natural gas and burned again. As a result, combustion efficiency is improved. Moreover, this technique is effective in reducing air pollution, particularly the emission of nitrogen oxides. Experimental data indicate that the use of twenty percent natural gas and eighty percent coal can reduce nitrogen oxides by as much as sixty percent. This remarkable reduction in nitrogen oxides is attributed to the following factors: (1) The clean-burning characteristics of natural gas; (2) Improved combustion of the gas-coal mixtures; and (3) The powerful affinity of natural gas (methane) for oxygen, which reduces the amount of oxygen available for combining with nitrogen.[5] The same combination of factors also reduces other pollutants.

A 1987 study by the American Gas Association estimates the potential select use of gas with coal-fired facilities could result in a demand of 200-700 billion cubic feet annually.[2] Mr. Ralbern Murray, Vice Chairman of Consolidated Natural Gas Company, noted that much larger amounts of natural gas may be used for this application in the mid-1990's, as coal-fired facilities become older and environmental requirements become stricter.[6]

Combined Cycle Power Generation

In contrast to repowering, which is designed around coal-fired equipment, regular combined cycle electric power generating facilities are completely new and utilize the most advanced technology. As a result, their efficiency is greater and the environmental benefits are optimum. Such a powerplant has the following components: (1) gas turbine; (2) electric generator; (3) heat recovery steam generator; (4) steam turbine; and (5) another electric generator. The gas turbine drives a generator, which produces electricity. The waste heat from the gas turbine is captured by the heat recovery steam generator, which feeds steam into a steam turbine, which drives a generator which produces electricity. This technology increases the overall energy efficiency of the powerplant. Currently available equipment achieves fuel efficiency of nearly 50 percent. Even higher efficiencies are considered likely in the future.[8] In contrast, the efficiency of conventional coal-fired powerplants is in the low 30's.

Combined cycle facilities have advantages similar to those already cited with repowering. Their capital and operating costs are lower than those of coal-fired plants. Typically, capital costs and non-fuel operating and maintenance expenses are about one-third those of coal units, which more than offsets the higher cost of natural gas compared with coal. Overall, gas-fired combined cycle generation has a levelized annual cost (i.e., cost over the lifetime of the facility) twenty to forty percent lower than coal-fired facilities.[1]

It takes about two years to design and construct a combined cycle power plant. Such equipment can be ordered on a modular basis, which gives the utility the flexibility to adjust its capital spending to actual trends in demand. Last but not least, combined cycle equipment has significant environmental advantages over other options.

As a result of these positive features, natural gas-fired combined cycle equipment is increasingly chosen by electric utilities to generate power in new facilities. Recent data indicate that about forty percent of planned utility-owned capacity additions of 100 megawatts and larger will be powered with natural gas.[1]

Purchasing Electricity from Outside Suppliers

The production of electricity by enterprises not owned by public utilities has been in a rapid growth trend. Non-utility generators (NUG's) have tripled capacity from 6,600 megawatts in 1985 to about 19,000 megawatts in 1988. As the following table indicates, such capacity is expected to double by 1997. The projections were made by the North American Electric Reliability Council.[9]

Non-Utility Generator Additions by Fuel Type, 1988-1997

Fuel Type	Megawatts	% of Total
Natural gas	3,029	15.0
Hydro	500	2.5
Coal	1,315	6.5
Geothermal	594	3.0
Wind	84	0.4
Solar	100	0.5
Refuse (solid waste)	458	2.3
Wood/wood wastes	346	1.7
Other or unknown*	13,702	68.1
Total	20,128	100.0

*Believed to be mostly gas-based capacity

Even if only half of the "other or unknown" capacity addition is fueled by natural gas, this fuel will account for about half of all the new electric generating capacity produced by non-utility generators.

Gas Can Help Meet the Electric Power Challenge

A combination of repowering old coal-fired facilities with gas, the select use of gas with coal, the construction of new combined cycle plants, and purchase of power from non-utility sources can meet anticipated electric power requirements in the mid-1990's and beyond. Natural gas will be the main fuel for all of these applications, provided it stays competitive in price. It is estimated that the use of natural gas for electricity generation will grow significantly over the next two decades. Gas use by electric utilities may reach levels nearly double the 2.8 quadrillion Btu currently used for this purpose. In addition, gas used in cogeneration plants and by independent power producers may grow by over 3 quadrillion Btu from the 1989 level of 1 Quad.[10] The gas industry is confident it can meet this demand.

Similarly, the equipment manufacturers have the facilities

and technological know-how to supply the demand by the electric utilities and non-utility generators. Speaking at a conference co-sponsored by the American Gas Association and the Edison Electric Insitute, Mr. Douglas M. Todd, Manager, Combined-Cycle Development, General Electric Company, stated, "The answer to the question: 'Can combined-cycle systems be relied upon to satisfy electricity shortfalls on a long-term basis?' is a definitive 'YES' providing the manufacturing capacity is scheduled in advance."[8]

Sources:

[1] *"Trends in Electric Generation Capacity and the Impacts on Natural Gas,"* Energy Analysis, *American Gas Assoication, January 23, 1989.*

[2] *"Gas Options to Offset Anticipated Electric Generation Shortages in the Mid-1990s," Ibid., May 21, 1987.*

[3] *"Status of the Electric Industry," John J. Kearney, Jr., Senior Vice President, Edison Electric Institute,* Efficient Electricity Generation with Natural Gas, *American Gas Association, 1988.*

[4] *"Repowering with Natural Gas—An Electricity Generation Option,"* Energy Analysis, *American Gas Association, October 17, 1986.*

[5] *"An Update on the Chaswick Gas/Coal Cofiring Experience," Steven Winberg, Director of Co-Firing, Consolidated Natural Gas Company and Bernhard P. Breen, President, Energy Systems Associates,* Efficient Electricity Generation with Natural Gas, *American Gas Association, 1988.*

[6] *Panel discussion on Developments in Natural Gas Cofiring Technology, remarks by Ralbern H. Murray, Vice Chairman, Consolidated Natural Gas Co., American Gas Association, June 6, 1988.*

[7] *"Conversion of Utility Gas Turbines to Combined Cycle Plants," E. Stephen Miliaras (presented at Energy Technology Conference XIII, Washington, D.C., March 1986).*

[8] *"Can the Combined Cycle Be Relied Upon to Satisfy Electricy Shortfalls on a Long-term Basis?" Douglas M. Todd, Manager, Combined Cycle Development, General Electric Co.,* Efficient Electricity Generation with Natural Gas, *American Gas Association, 1988.*

[9] 1988 Reliability Assessment, *North American Electric Reliability Council, Princeton, New Jersey, September 1988.*

[10] *"The A.G.A TERA 1989 Mid-Year Base Case," American Gas Association, August 16, 1989 (draft).*

17. Promising Prospects for Gas in Commercial Markets

Natural gas is the fuel of choice for commercial space heating, hot water production, cooking, and drying. New technologies, such as gas cooling and cogeneration, will find increased applications in commercial markets. In 1987, 4.0 million commercial accounts used 2.5 quadrillion Btu of gas.[1] It is likely that these numbers will grow significantly over the next several years.

Electricity is the main competitor of gas in commercial markets. In 1988, the cost of electricity to commercial customers averaged $21.49 per million Btu; the cost of gas for the same amount of energy was $4.71.[2] This four-to-one price advantage of gas over electricity provides strong inducements for commercial energy users to switch to gas.

The cost of energy is a major factor in overall operating expenses of commercial eneterprises. These costs can be reduced significantly by carefully analyzing current energy consumption patterns and exploring alternative options. By replacing electricity with gas for space heating, cooling, and hot water production, substantial savings can be achieved. As an approximate guideline, investments in energy savings make good economic sense if the annual return is twenty percent or more. In some cases, notably hot water production, savings may range from thirty to fifty percent annually. In the aggregate, commercial enterprises could probably save several hundred million dollars a year if they used gas instead of electricity for producing hot water.

Electric air conditioning systems involve large energy expenses, particularly in view of the fact that they are used primariliy in summer months, when electric rates are at their peak. Efficient gas cooling equipment is now available which can reduce these costs substantially.

Cogeneration is another option worthy of serious attention by commercial energy users. This technology produces both electricity and thermal energy (heat). If heat requirements are high, cogeneration is an attractive choice. Hospitals, hotels, and office buildings are particularly well suited for the use of gas-fired cogeneration. A wide range of packaged gas-fired cogeneration systems is offered in the marketplace.

Many older commercial buildings use fuel oil for space heating. If an oil boiler or furnace is more than twenty years old, it is likely to be very inefficient. Typically, it may waste as much as half or more of the energy consumed. By replacing such an old heating unit with a new high-efficient gas boiler or furnace, large savings can be realized. Such an investment often pays for itself in less than four years. Recent rules by the Environmental Protection Agency make it necessary for owners of underground oil storage tanks to guard against leaks. To comply with these rules, owners may need to install leak detection devices and to double-line the tanks.[3] These expenses can be avoided by switching to gas.

Supermarkets and other enterprises handling large quantities of food should consider gas-fired desiccant dehumidification systems which offer improved air quality, precise humidity control, and reduced electric demand for cooling.[3] This technology provides an optimum environment for fresh produce.

Most electric air conditioning systems use chlorofluorocarbons (CFC's) which emit pollutants harmful to the earth's protective ozone layer. Gas absorption air conditioning does not involve the use of CFCs.[3]

Many commercial enterprises make extensive use of vehicles. These fleets could achieve significant savings and improve environmental performance by switching to compressed natural gas (CNG).

In the following sections, information is provided on energy matters relating to schools, hospitals, large buildings, and commercial kitchens. Much of this information is also useful to other commercial enterprises.

Schools

Schools are major users of energy. In fact, energy and utility services account for 30-35 percent of the operating and maintenance budgets of educational facilities.[4] A wide range of gas equipment is available to help schools save money on their energy needs. Gas equipment requires minimal maintenance and is easy to install.

Tens of thousands of schools throughout the country have antiquated heating systems, some using oil, that should be replaced. If the present heating system is more than twenty years old, its efficiency is likely to be very low, resulting in high fuel bills. By switching to high-efficient gas heating equipment, large savings can be realized. Similarly, the replacement of an old water heater with a new efficient gas unit can be a very profitable investment. It makes good sense to think of energy savings as investment opportunities. The annual returns on such investments are usually far greater than those available form alternative options.

Many schools maintain fleets of buses to transport students between their homes and the school By converting these buses to compressed natural gas (CNG), considerable fuel savings can be achieved. If a school is in the market for new buses, it may want to consider vehicles with dedicated natural gas engines, which are now available. In addition to cost savings and improved air quality, such vehicles could be used for educational purposes. They give students a living demonstration of how natural gas can help clean the environment.

Hospitals

Energy accounts for about 15 percent of the operating budget of health care facilities.[5] Hospitals are energy-intensive enterprises. They have special requirements which add considerably to energy use and costs. Federal and local regulations mandate that hospitals maintain strict tolerances in temperature and humidity levels in order to achieve optimum patient comfort. Moreover, hospitals use procedures that exhaust large quantities of conditioned air to the outside to minimize risks of contamination. Space heating and cooling loads are much higher in hospitals than in most other places.

Hospitals are large users of hot water and steam, which are used in laundries, kitchens, and sterilizing facilities. The disposal of hazardous medical waste through incineration is of increased importance to hospitals.

Natural gas can play a helpful role in all of these applications. It is the lowest cost fuel for space heating and cooling. It is the best fuel for hot water production. It can help minimize pollution from incineration.

Because of their energy-intensive nature, with large requirements of both electricity and heat energy, hospitals are well suited for cogeneration. This option for generating their own electricity and heat energy should be seriously investigated by most hospitals. In addition to making possible substantial cost savings, this technology has the advantage of making the hospital largely self-sufficient in taking care of its energy needs.

Many hospitals use energy equipment that is old and inefficient. In many cases, large savings can be realized by switching to new high-efficient gas equipment for space heating and cooling, and hot water production. Cogeneration is particularly well suited for hospitals that want to overhaul their whole energy system, as well as for new hospital construction.

Large Buildings

Office buildings, hotels and motels, and multi-family apartment buildings are large energy users.[6]

A wide range of efficient gas furnaces is available to serve the needs of these buildings. Choices include central heating systems as well as units for individual apartments and rooms. Similarly, gas water heaters range in capacities small enough for an efficiency apartment to a size sufficient to meet the needs of the largest building. Gas dryers also have a wide range, from small appliances for use within an apartment, to large units for installation in laundry rooms.

The availability of packaged gas-fired cogeneration systems, in a wide range of sizes, may be of particular interest to the managers of large commercial buildings. Cogeneration becomes especially attractive to buildings that use a great deal of thermal energy. For example, hotels with convention facilities, multiple restaurants, in-house laundry, and central heating and air conditioning systems are good prospects for cogeneration.

Commercial Kitchens

Natural gas is the favorite fuel of most commercial kitchens, which use more than 300 billion cubic feet annually.[7] Chefs prefer gas because it allows maximum control over the flame. Kitchen equipment fueled with gas is reliable and easy to maintain. There is a wide range of choices of equipment. Moreover, gas is considerably cheaper than electricity. A 1982 study by the University of Minnesota showed that gas is 30-60 percent cheaper than electricity for the most commonly used cooking procedures.[7]

A great many new products have been introduced for commercial kitchens in recent years. Here are some noteworthy examples:

(1) Convection ovens with fans which circulate heated air and speed up the cooking process.

(2) Roast and hold convection ovens that roast at low temperatures and hold at even lower temperatures. This

procedure improves the quality of roasted meats, reduces shrinkage, and saves energy.

(3) Electronic ignition systems, which eliminate pilot lights and reduce fuel consumption.

(4) Solid state temperature controls which are more accurate and more dependable than old-style controls.

(5) Tilting braising pans, which increase production and lower fuel costs.

(6) Infra-red griddles, which provide greater heat and improved energy efficiency.

(7) Infra-red fryers, which reduce fuel consumption from 145 MBtu to 80 MBtu.

(8) Conveyor type infra-red broilers, which broil both sides at once, saving time and labor.

(9) Pressureless convection steamers, which are particularly useful for quickly defrosting frozen foods and for cooking delicate products like spinach. The doors of such steamers can be opened to inspect the product without interfering with the steaming process.

(10) A variety of high-efficient gas water heaters, which produce hot water at substantially less cost than electric units.

Whenever they need new cooking equipment, kitchen managers and chefs should check for the latest offerings by gas equipment dealers. New products are likely to be available to serve their needs more effectively at lower operating costs.

To sum up, the prospects for the increased use of natural gas look promising in many areas of the commercial market.

Sources:

[1] Gas Facts: 1987 Data, *American Gas Association, 1988.*

[2] Monthly Energy Report: April 1989, *Department of Energy, Energy Information Administration, July 1989.*

[3] Gas Industry Five-Year Strategic Marketing Plan, 1989-1993, *American Gas Association, December 1988.*

[4] *"The School Official's Guide to Natural Gas," American Gas Association, 1989.*

[5] Efficient Use of Natural Gas in Hospitals, Nursing Homes, and Health Care Facilities, *American Gas Association, 1985.*

[6] Efficient Use of Natural Gas in Office Buildings, *A.G.A., 1985.*

Efficient Use of Natural Gas in Hotels and Motels, *A.G.A., 1984.*

Efficient Use of Natural Gas in Multi-Family Buildings, *A.G.A., 1984.*

[7] Efficient Use of Natural Gas in Commercial Kitchens, *A.G.A., 1986.*

18. Energy-efficient Industry, A Good Gas Market

Prior to 1973, U.S. industry made maximum use of cheap energy sources to minimize production costs. Very little attention was paid to energy efficiency. This orientation underwent a fundamental transformation after the oil price escalations of the 1970's and the resulting price increases in all forms of energy. Between 1973 and 1986, the U.S. industrial sector *reduced* energy consumption by 23 percent, while *increasing* production by 33 percent.

The following table shows energy consumption by U.S. industry between 1973 and 1986. The industrial category includes manufacturing, agriculture, construction, and mining.

Table 1: U.S. Industrial Energy Consumption[1]
(Quadrillion Btu)

1973	31.53	1980	30.61
1974	30.69	1981	29.24
1975	28.40	1982	26.14
1976	30.24	1983	25.74
1977	31.08	1984	27.72
1978	31.41	1985	27.03
1979	32.62	1986	26.45

Energy consumption trends were closely linked to the oil price escalations of 1973 and 1979. A modest decline occurred between 1973 and 1975, primarily as a result of the recession in 1974-75. In the following years, energy consumption increased in line with economic growth. The response to the second oil price escalation (in 1979) was different. The decline in energy consumption between 1979 and 1982 (a recession year) was much sharper than the earlier experience, plummeting from 32.62 Quads in 1979 ato 26.14 Quads in 1982. Moreover, while industrial production rose substantially between 1983 and 1986, energy consumption by industry failed to keep pace and was near its low by 1986. Industrial production and energy consumption had become decoupled.

Table 2 presents energy purchases by the manufacturing sector, which accounted for a large part of the decline in energy consumption.

Table 2: Manufacturing Sector
Energy Purchase Requirements, 1973-1986[2]
(Quadrillion Btu)

Year	Natural Gas	Oil	Coal	Electricity	Total
1973	6.60	1.82	3.87	2.07	14.28
1974	6.70	1.74	3.00	2.10	13.55
1975	5.93	1.73	2.39	2.04	12.08
1976	5.98	1.98	2.63	2.18	12.78
1977	5.68	2.12	2.82	2.26	12.89
1978	5.59	2.12	2.89	2.31	12.90
1979	5.98	1.56	2.99	2.34	12.87
1980	5.69	1.16	2.85	2.25	11.95
1981	5.51	0.95	2.79	2.27	11.56
1982	4.53	0.92	2.94	2.05	10.44
1983	4.26	0.71	2.94	2.21	10.12
1984	4.79	0.78	3.03	2.39	10.99
1985	4.62	0.71	3.19	2.29	10.80

The biggest percentage decline in energy use by manufacturers took place in oil, which dropped from 1.32 Quads in 1973 to 0.71 Quads in 1985, a loss of 61 percent. Natural gas consumption declined 29 percent in the same period. Coal experienced only a modest reduction, while electricity consumption increased by ten percent.

Price is the most important determinant of energy usage in the manufacturing sector. Table 3 examines price trends in the period 1973-1986. Prices are based on cost per million Btu, in constant 1982 dollars.

Table 3: U.S. Industrial Energy Prices[3]

Year	Natural Gas	Dist. Fuel Oil	Resid. Fuel Oil	Coal	Electricity
1973	1.01	1.86	1.47	1.25	7.39
1974	1.24	3.74	3.37	2.26	9.17
1975	1.60	3.76	3.22	2.55	10.24
1976	1.92	3.77	3.01	2.38	10.27
1977	2.20	3.98	3.19	2.32	10.89
1978	2.30	3.91	2.94	2.40	11.33
1979	2.48	4.90	3.52	2.21	11.37
1980	2.93	6.46	4.32	2.18	12.63
1981	3.27	6.95	4.77	2.19	13.37
1982	3.79	6.63	4.45	2.09	14.51
1983	3.94	5.96	4.21	1.85	14.01
1984	3.80	5.88	4.38	1.77	13.66
1985	3.46	5.37	3.83	1.70	13.54
1986	2.71	4.18	2.09	N.A.	11.90

Oil products showed the greatest price volatility. Distillate fuel oil rose from $1.86 per million Btu in 1973 to $6.95 in 1981. It subsequently declined to $4.18 in 1986. Residual fuel oil jumped from $1.47 in 1973 to $4.77 in 1981. By 1986, it had dropped to $2.09. Natural gas prices more than tripled between 1973 and 1981 (from $1.01 to $3.27). Moreover, gas prices continued to rise in 1982 and 1983, while other energy prices were declining. This anomaly, which was partly caused by regulated prices, proved to be very costly to the gas industry, which lost 1.25 Quads of industrial sales during those two years. Since 1985 gas prices have declined.

Price increases in coal were more moderate than those in other energy sources. Coal prices also showed much less volatility than oil or natural gas. As a result, demand for coal remained relatively constant throughout the period 1973-1986. Manufacturers like fuels that minimze price surprises and volatility.

The cost of electricity is largely determined by prices for fuels and capital (interest rates). Electric utilities generally use the lowest cost fuels they can obtain, with primary emphasis on coal. The relative price stability of coal during much of the period helped keep down the cost of electricity. Similarly, the electric generating facilities that were built prior to 1973 were financed at low interest rates, which also helped to keep down costs. However, interest rates have risen sharply since 1973. As a result, electric utilities are losing a major cost advantage as they are faced with the problem of replacing old facilities with new ones.

As a general observation, electric utilities are not eager to construct new generating facilities primarily to serve industrial markets, which yield the lowest prices of any customer category. Therefore, industrial users of electricity are increasingly turning to cogeneration. This technology enables them to use the same fuel for producing both electricity and heat, thereby optimizing energy efficiency and reducing costs. Natural gas is in an advantageous position for fueling cogeneration systems. Capital and operating costs of gas-fired cogeneration are lower than those of competing systems utilizing coal.[4] Similarly, gas avoids the necessity for heavy investments in pollution control equipment. The trend toward increased use of cogeneration by many industrial customers favors natural gas.

The biggest industrial users of natural gas are: (1) chemicals; (2) petroleum refining; (3) food; (4) stone, clay, glass; (5) steel; and (6) paper. The next table shows gas consumption trends in those industries.

Table 4: Large Industrial Users of Natural Gas
(trillion Btu)

Year	Chemicals[5]	Petroleum Refining[6]	Food[7]	Stone, Clay, Glass[8]	Steel[9]	Paper[10]
1973	1,814		472	696	642	
1974	1,814		472	690	670	
1975	1,634	946	451	585	577	338
1976	1,754	928	448	599	595	324
1977	1,579	914	437	534	568	313
1978	1,568	804	427	550	593	315
1979	1,689	840	455	549	544	334
1980	1,922	829	485	498	560	381
1981	1,812	651	471	464	613	400
1982	1,344	591	465	381	411	340
1983	1,320	573	472	393	379	314
1984	1,392	573	476	397	391	329
1985	1,456	488	459	397	348	306
1986		582			309	288

The chemical industry is the largest industrial user of natural gas. Ammonia fertilizer, which uses natural gas as feedstock and fuel, accounts for the biggest share of the total. Because the chemical industry is a heavy user of steam, it is in a good position to take advantage of cogeneration. The growth of cogeneration in the chemical industry has been spectacular, rising from 3,000 megawatts in 1980 to an estimated 13,600 megawatts in 1986.[11] About 600 Bcf of natural gas is used for cogeneration in the chemical industry. Additional growth in this application is likely over the next several years.

Plastics is another growth area for natural gas. Natural gas use in plastics was estimated at 190 Bcf in 1985.[12]

It is likely that gas consumption by the chemical industry will follow closely overall production trends by that industry.

Most of the decline in natural gas use in petroleum refining may be attributed to the substitution of "still gas," also called "refinery gas," which is a byproduct of petroleum refining. In 1986, natural gas use increased significantly; this trend has continued in 1987, when natural gas consumption by petroleum refiners reach 609 Bcf.[13]

Gas consumption by the food industry has remained stable during the period 1973-1985. The industry achieved considerable energy efficiency improvements by using the same amount of gas energy for a larger amount of total output. However, gas has remained the dominant fuel for this industry, which places a premium on cleanliness. Since 1980, gas has accounted for more than fifty percent of total energy use by the food industry.[14]

The outlook is favorable for increased use of natural gas by the food industry. Small-scale cogeneration systems have a promising future in this industry, which makes heavy use of thermal energy as well as electricity. According to recent estimates, active cogeneration projects in the food industry have a combined capacity of 1,600 megawatts.[14]

The food industry also uses large amounts of refrigeration to keep its products fresh. Gas cooling and refrigeration systems have a big potential market in this field, which has hitherto been dominated by electricity.[14]

The use of natural gas in the manufacture of stone, clay, and glass dropped from 696 Bcf in 1973 to 397 Bcf in 1985. Fuel switching from gas to coal, particularly in the cement, clay, and lime industries, accounted for much of the decline, along with improved efficiency. The use of natural gas in glass production has increased from 36 percent of total fuels in 1971 to 59 percent in 1985.[15]

Under the sponsorship of the Gas Research Institute, an advanced gas-fired cement production technology is being developed. This process reduces energy consumption by 35 percent. Capital and operating costs are considerably lower than those applicable to coal technology. This development may help gas regain some of the market it lost to coal in the manufacture of cement.[15]

The decline in the use of natural gas for steel production was caused by a combination of lower output of steel and greater efficiency. In 1987, gas consumption by the steel industry increased to 396 Btu, reflecting larger steel production. Gas demand by this industry is likely to continue reflecting changes in output, which is heavily influenced by cyclical factors and by imports.

The most important raw material of the paper industry is wood. For thousands of years, wood was the main source of fuel for most of mankind. The energy price escalations of the 1970's induced the paper industry to make increased use of self-generated and residual fuels derived from wood to meet its energy requirements. The use of such fuels increased from 849 trillion Btus in 1972 to 1,303 TBtus in 1986. In the same interval, total purchased energy declined from 1,245 TBtus to 970 TBtus. Natural gas was one of the purchased fuels sharing this decline.[17]

The paper industry is the largest user of cogeneration, with 7,500 megawatts on line in 1986. Most of this cogeneration was fueled with spent liquor (a byproduct of paper manufacture), wood wastes, and coal. Gas accounted for 12 percent of the total. Cogeneration fueled with natural gas has lower capital and operating costs than those using coal or spent liquor. These cost advantages present opportunities in some cases for new installations or for replacements of older facilities.

Since the 1970's, energy efficiency has become a major consideration for U.S manufacturers. Great strides have been made in reducing energy consumption per unit of output. While natural gas initially suffered loss of market share as a result of this trend, it is experiencing renewed strenngth as a result of premium applications, such as industrial cogeneration and other sophisticated technologies. The increased emphasis on environmental cleanliness will also help natural gas gain ground in industrial markets.

Sources:

[1] Monthly Energy Review, *U.S. Department of Energy, Washington, D.C.*

[2] Survey of Manufacturers, *Department of Commerce, Washington, D.C. For the period 1981-1984, the Commerce Department did not publish data on energy purchases of manufacturers. The data for this period were estimated by the staff of the American Gas Association.*

[3] *"Seper Data Base," U.S. Department of Energy, Washington, D.C.*

[4] *"An Economic and Enviromental Comparison of Natural Gas nad Coal Use for Large-Scale Industrial Cogeneration,"* Energy Analysis, *American Gas Association, October 26, 1984. See also 1989 Update, June 9, 1989.*

[5] *Gas Requirements Committee of the American Gas Association.*

[6] Basic Petroleum Data Book, *American Petroleum Institute, 1988. Also,* Petroleum Annuals, *1975-1980 and* Petroleum Supply Annuals, *1981-1986, Energy Information Administration;* Natural Gas Trends, *1987-88 Edition, Cambridge Energy Research Associates, Cambridge, MA.*

[7] Manufacturing Energy Consumption Survey, *Energy Information Administration;* Annual Survey of Manufacturers, *Department of Commerce, for the period 1974-1981; American Gas Association estimates for the years 1973 and 1982-1984.*

[8] Annual Survey of Manufacturers, *Department of Commerce, 1973-1981. Data for 1982-1984 are estimates by the Gas Requirements Committee of the American Gas Association; Data for 1985 were from* Manufacturing Energy Consumption Survey, *Energy Information Adminstration.*

[9] Annual Statistical Reports, *American Iron & Steel Institute.*

[10] U.S. Pulp and Paperboard Industry Energy Use Survey, 1975-1986, *American Paper Institute.*

[11] Profile of Cogeneration and Small Power Generation Markets, *Hagler, Bailly & Company, Washington, D.C., distributed in cooperation with Utilty Data Institute, Washington, D.C., August 1987.*

[12] *Section on Chemicals, by James Pavle,* Industrial Sector Energy Consumption and the Outlook for Natural Gas, *Planning & Analysis Group, American Gas Association, December 1988.*

[13] *Section on Petroleum Refining, by Bruce B. Henning, ibid.*
[14] *Section on Food and Kindred Products, by Christine Swanson, ibid.*

[15] *Section on Stone, Clay and Glass, by Russell Tucker, ibid. The author obtained additional information about the advanced gas-fired cement production technology from Mr. Gregory L. Ridderbusch of the Gas Research Institute.*

[16] *Section on Steel, by Paul F. McArdle, ibid.*

[17] *Section on Paper, by Paul F. McArdle, ibid.*

19. The Cleanest and Safest Fuel for Vehicles

Vehicles fueled with compressed natural gas (CNG) have the following advantages:

(1) They minimize air pollution.
(2) Their engines last longer and require less maintenance.
(3) They have lower fuel costs.
(4) They have a better safety record.
(5) They strengthen U.S. energy security.

Natural gas can play a key role in reducing air pollution prolems caused by vehicles. A study by the California Air Resources Board found that natural gas significantly lowered the emission of pollutants that cause smog (groound-level ozone). The following table provides details.

Air Pollution Emissions[1]
(grams per mile)

Pollutant	Gasoline	Natural Gas
Carbon monoxide	1.40	0.10
Nitrogen oxides	0.66	0.44
Exhaust reactive organic gases	0.35	0.19
Evaporation reactive organic gases	0.04	None

Princeton University's Department of Mechanical and Aerospace Engineering also found natural gas to be a much cleaner-burning fuel than gasoline. The study explained that the primary advantages of natural gas resulted from its low ratio of carbon to hydrogen and from its abilitiy to mix more thoroughly with air than gasoline. Carbon monoxide is formed when a fuel does not mix well with oxygen. Gasoline generates relatively large amounts of carbon monoxide, while natural gas mixes so well with air that carbon monoxide is minimized. The Princeton University study attributed the relatively high levels of reactive hydrocarbons in gasoline to the presence of olefins and aromatic compounds that are used to raise the octane of unleaded gasoline. These compounds are not present in natural gas, which has an octane rating of 130 without any additives. The thorough mixing of natural gas with air also reduces nitrogen oxide emissions in comparison with gasoline. The Princeton University study gives natural gas a high rating for environmental cleanliness in vehicles.[2]

The clean-burning characteristics of natural gas protect the engine. As a result, natural gas engines have a longer useful life and require less maintenance than gasoline engines.

While the price of natural gas varies, depending on location and utility policy, in general the cost of natural gas is less than the cost of gasoline. In 1989, utilities charged operators of fleet vehicles prices ranging from 40 to 75 cents per gallon (on an energy equivalent basis).[3]

Natural gas vehicles have a better safety record than their gasoline counterparts. In a survey involving natural gas vehicles used in regular operations by utility personnel driving 434 million miles, the collision rate was about the same as the national average on such mileage. However, the number of injuries was 84 percent lower in natural gas vehicles than in those using gasoline (10.1 injuries per 100 million miles in natural gas vehicles versus 63.7 injuries in gasoline vehicles). Moreover, there were no deaths in natural gas vehicles, while the norm for gasoline vehicles would have been 2.47 deaths per 100 million miles, or about 11 for the 434 million miles driven by the natural gas vehicles.[4]

From a fire standpoint, natural gas as a vehicular fuel is significantly safer than gasoline. Compressed natural gas is stored in a cylinder constructed of steel, aluminum, or fiberglass. This container is so strong that it is unlikely to break open as a result of a collision. Moreover, even if it were ripped open, the gas, being lighter than air and having a high ignition point, would not ignite, but would simply escape into the air. In contrast, the gasoline tank can easily break open during a collision. The gasoline puddles to the ground and often ignites. It may well be a major cause of injuries and deaths. Currently available statistics for gasoline vehicles do not single out injuries or deaths caused by fires resulting from gasoline ignition during accidents. In the case of the natural gas vehicles, there were only two fires attributable to the compressed natural gas system during 434 million miles of driving.[4]

This subject is sufficiently important to justify additional studies by independent research organizations and/or government agencies.

The U.S. has plentiful domestic natural gas resources. The only imports of natural gas come from Canada; they account for less than ten percent of U.S. consumption. In contrast, oil imports amount to more than forty percent of U.S. requirements. Moreover, dependence on oil imports is growing. The oil price escalations and supply disruptions during the past sixteen years provide ample evidence that excessive dependence on foreign sources of oil involves high risks and potential harm to the economy. Natural gas vehicles are a sound investment in long-term energy security.

Currently, there are about 30,000 vehicles in the U.S. fueled with natural gas. On a worldwide basis, the number of gas-fueled vehicles exceeds 700,000 and is growing rapidly. The Soviet Union has set a goal of converting one million vehicles to natural gas. Other major users of natural gas vehicles include Italy, New Zealand, Canada, and Argentina.[5]

For vehicular applications, natural gas is compressed and stored in metal or fiberglass cylinders. The gas is connected by a fuel line to the engine. Hitherto, the practice has been to utilize gasoline or diesel engines, which have the capability of running on either liquid or gaseous fuels. This dual-fuel capability has the advantage of increasing the range of the vehicle and allowing it to fill up on either fuel. However, this approach has the drawback of not making optimum use of the energy contained in compressed natural gas.

Methane, which is the principal ingredient of natural gas, has an octane rating of 130. Gasoline's octane rating is less than 100. The gasoline engine is built for optimum functioning with gasoline, not with compressed natural gas.

If natural gas is to realize its potential as a major contributor to a cleaner environment and greater energy security, it is essential to develop engines specifically dedicated to this fuel. The following companies are currently engaged in developing dedicated natural gas engines and equipment, and/or have gained experience in that field.[3]

(1) Cummins Engine Corporation has developed a four-stroke diesel engine modified for natural gas. These engines are currently being field-tested and are expected to go into production in 1990. They are designed for use on mass transit buses.

(2) MTN Energy Systems of Los Angeles has developed the Thunder engine, which is being adapted to natural gas use with support from the Gas Research Institute. This high-performance engine, which was originally designed for aircraft use, utilizes an aluminum block, which cuts its weight in half compared with a conventional diesel engine. The weight advantage is valuable in connection with natural gas vehicles, which carry gas cylinders that may be fairly heavy.

(3)Hercules Engine Inc has begun developmental work on a diesel engine modified for natural gas to be used for medium duty delivery trucks. Field test evaluation is planned for 1990.

(4) The Detroit Diesel Corporation is working on a two-stroke diesel engine dedicated to using compressed natural gas (CNG). A retrofit version of this engine for CNG use in existing vehicles is being developed by Stewart and Stevenson.

(5) Bus Industries of America and Flexible Corporation, major producers of transit buses, have produced prototypes of urban transit buses utilizing compressed natural gas. Two

buses built by Bus Industries of America were purchased by the Brooklyn Union Gas Company. These buses have been made available to the New York City Department of Transportation for field testing, which is currently underway.[6]

(6) The Ford Motor Company has equipped 27 Ranger pickup trucks with dedicated natural gas engines. The company's engineers expressed confidence that they could mass-produce such engines when market conditions justify the investments involved.[3]

(7) General Motors is testing a modified version of their 4.3 liter gasoline engine to run on natural gas. This engine is designed for use on vans and pickup trucks. Large-scale production of these natural gas engines would be feasible if demand were sufficient.[3]

(8) The GSM Taxi Ltd company of St. Laurent, Canada, has produced factory-built natural gas vehicles for the utility market. The Natural Gas Vehicle Coalition is supporting the marketing of these vehicles to gas utilities and their customers.[3]

The 1,150,000 miles of gas distribution pipelines and mains provide a sound foundation for an infrastructure of filling stations. However, gas filling stations are quite different from those dispensing liquid fuels. The heart of the natural gas filling station is a compressor, which compresses the gas for use in vehicles. In contrast, filling stations for gasoline and diesel fuel use pumps to fill vehicles. Several factors enter into the cost of building natural gas filling stations, including the price of the compressor and related equipment, site acquisition and preparation, and link-up to gas mains. As an approximate estimate, the cost of a filling station is likely to range from $100,000 and up. It is evident that the establishment of a national network of natural gas filling stations would require large investments.

At the present time, about 275 natural gas filling stations in the U.S. are available to service vehicles with CNG. Most of these stations are situated on the premises of gas utilities. Some are at locations owned by fleet operators.[8]

In the U.S. there are approximately 75,000 fleets that own

an estimated 16 million vehicles.[7] These fleets are the best potential markets for natural gas vehicles. Many of them have the facilities for building their own filling stations. They make heavy use of vehicles, frequently within a metropolitan area, which allows for optimizing the cost advantages of natural gas over competing fuels.

Federal, state and local governments are the biggest users of fleet vehicles. Altogether, they operate 2.6 million vehicles, including buses, trucks, police cars, vans, and passenger vehicles.[7] By shifting even a portion of their vehicular purchases to natural gas, governments could play a major role in facilitating the emergence of an economically viable natural gas vehicle industry.

Governments generally set high standards for performance on vehicles they purchase. They are looked upon as leaders by the private sector. If governments switched to natural gas vehicles on a large scale, it is likely that many private fleet operators would also order such vehicles. This process would stimulate the construction of natural gas filling stations, which would encourage additional conversions to natural gas vehicles. This upsurge in demand for natural gas vehicles would induce manufacturers to mass-produce them, which would lower prices and make such vehicles competitive with conventional cars, trucks, and buses.

The federal government has taken some steps to encourage the use of alternative fuels, including natural gas. The Alternative Motor Fuels Act of 1988 establishes incentives for car manufacturers to produce alternative fuel vehicles and provides funds for the purchase of such vehicles by the federal government. The legislation allows vehicle manufacturers to earn credits toward attaining "corporate average fuel economy" (CAFE) standards by producing vehicles that operate on alternative fuels. This legislation also authorizes the expenditure of $16 million over a four-year period to purchase alternatively fueled cars and trucks. In addition, $2 million were authorized for demonstrating alternatively fueled buses.[8]

In September 1988, the Urban Mass Transportation Authority

(UMTA) of the U.S. Department of Transportation announced a $46.8 million program to promote the purchase of alternative fueled transit buses. The program provided about $35 million federal funds, to be matched by 25 percent private, state, or local funds. As of July 1989, UMTA had received orders for about 420 natural gas buses from transit authorities.[3]

Several state governments, including those of Texas, California, and Arizona, have taken action to reduce air pollution by encouraging the use of non-polluting fuels. The State of Texas has passed legislation that empowers the Texas Air Control Board to take decisive steps to bring metropolitan areas into compliance with federal ambient air quality standards. This legislation makes specific mention of natural gas as an appropriate means for helping to accomplish this goal.[9] This legislation enjoyed wide support among leading Texans and the mass media in that state. This broadly based stand in favor of natural gas for environmental reasons is particularly significant for a state that is the nation's leading oil and gas producer. It seems to indicate that Texas government officials and a broad segment of the private sector, including energy producers, have reached the conclusion that their own interests are better served with natural gas than with oil.

To sum up, natural gas vehicles can play an important role in helping to reduce air pollution and to strengthen energy security. Substantial progress has been made in developing the technologies for natural gas vehicles. Governments have taken the first steps in the direction of encouraging alternative fuels, including natural gas, for environmental reasons. The natural gas industry is actively involved in supporting the effort to expand the use of such vehicles. The outlook is favorable for the rapid growth of natural gas vehicles in coming years.

Sources:

[1] *"California Air Resources Board Definition of a Low Emission Motor Vehicle in Compliance with the Mandates of Health and Safety Code Section 39037.05," May 19, 1989.*

[2] *"Understanding Emissions Levels from Vehicle Engines Fueled with Gaseous Fuels," Princeton University, Department of Mechanical and Aerospace Engineering, Spring 1989.*

[3] *"The Future of Natural Gas Vehicles," by Jeffrey Seisler. The Natural Gas Vehicle Coalition, Washington, D.C. March 1989.*

[4] *"Natural Gas Vehicle Safety Survey—An Update,"* Issues Brief 1987-No. 6, *American Gas Association, June 26, 1987.*

[5] *"Natural Gas Vehicles: The International Experience," American Gas Association, May 13, 1988.*

[6] *For additional information, contact the Brooklyn Union Gas Company, Natural Gas Vehicle Section, 166 Montague Street, Brooklyn, New York 11201.*

[7] *"Toward Commercialization of NGV's," by Jeffrey Seisler. American Gas Association, March 1984.*

[8] *"Alternative Motor Fuels Act of 1988,"* Legislative Analysis, *American Gas Association, 4L-1988.*

[9] *"Senate Bill 740 (1989), Henderson/Cain—Alternative Fuels Program." "Senate Bill 769 (1989), Caperton/Erlanca—Clean Air Act Amendments." Bill summaries prepared by the Office of the Commissioner, General Land Office, State of Texas (Mr. Garry Mauro). Stephen F. Austin Building, Austin Texas 78701.*

PART III
GAS ADVANTAGES

20. Environmental Benefits

Natural gas is the cleanest of all fossil fuels. Its production involves minimal disturbance of the surroundings. Cleaning of the gas *before* it is put into pipelines removes most pollutants. Emissions resulting from the combustion process are considerably less than those produced by other fossil fuels. Natural gas is worthy of support by all those concerned with the environment.

Drilling for natural gas is basically a clean operation, involving very little permanent disturbance of the surrounding area. Similar comments are applicable to gas production procedures. There are no unsightly structures, nor is there any large-scale removal of plants or soil. Once production has been completed, the drilling installations are removed and the environment is quickly restored to its original state.

To meet pipeline standards, virtually all pollutants are removed from the natural gas *before* it is allowed entry. By the time it reaches the consumer, gas is as clean as any fossil fuel can be. In contrast, with most other fossil fuels the attempt is made to remove pollutants *after* they are burned, a procedure which is inherently less effective.

The burning of natural gas, which is primarily methane, is very simple and clean. In the principal reaction, one methane molecule (CH_4) combines with two oxygen molecules (O_2) to form two water molecules ($2H_2O$) and one carbon dioxide molecule (CO_2). The chemical shorthand for this transaction is as follows: $CH_4 + 2O_2 \rightarrow 2H_2O + CO_2$. The water molecules are in the form of vapor, which will eventually return to the ground in the form of rain or dew. Carbon dioxide serves as a food for plants.

The cleanliness of gas is illustrated by the following table, which compares gas with oil and coal. The data for this table are based on information provided by the Environmental Protection Agency,[1] the U.S. Department of Energy,[2] and Hittman Associates, Inc.[3]

Pounds of Air Pollution per Billion Btu

	Gas	Oil	Coal
Sulfur oxides	0.6	830-920	660-4,390
Particulates	5-15	140-720	60-9,440
Carbon monoxide	17-20	40	44-88
Hydrocarbons	1-8	7	13-44
Nitrogen oxides	80-700	130-760	670-2,440

A comment about nitrogen oxides may be in order. Nitrogen is a constituent of air and will form compounds with oxygen (nitrogen oxides) whenever combustion takes place in the presence of air. As the above table shows, even in this category gas has significant environmental advantages over other fossil fuels, particularly coal.

The environmental cleanliness of natural gas is especially valuable in reducing excessive ground level (tropospheric) ozone, which will be discussed in the following section.

Gas Helps Minimize Ozone Pollution

Excessive ground level ozone has become a major air pollution problem in areas containing more than one-third of the U.S. population. Nearly every major city fails to meet the ozone standards set by the Environmental Protection Agency.[4,5] Because ozone formation is related to temperature and sunlight, most regions of the U.S. suffer from excessive ozone during the summer months. However, in the southwestern section and in Florida the problem may exist throughout most of the year.[5,6]

Ozone is an oxygen molecule with three atoms (O_3) instead of the two (O_2) most prevalent in the air. It is an unstable oxydizing agent with a pungent odor. It has a bluish color (normal oxygen is colorless). Most ozone formation takes place at temperatures above 65° Fahrenheit. Human beings are adversely affected by ozone, which irritates the nose and eyes and aggravates respiratory illnesses.[7,8] Crops and many materials are damaged by ozone.

Chemical reactions among nitrogen oxides, reactive hydrocarbons (such as gasoline vapors and chemical solvents), and carbon monozide in the presence of sunlight are the primary sources of ozone. Vehicles powered with gasoline or diesel fuel are major contributors to ozone formation. In addition, industrial processes and coal-fired facilities add to the chemicals that produce ozone.[9]

The main sources of ozone pollution may be summarized as follows:

(1) In terms of quantity, carbon monoxide heads the list, with an estimated 60.9 million metric tons emitted into the air annually. Vehicles are the source of more than two-thirds of this pollutant.[5]

(2) An estimated 19.5 million metric tons of reactive hydrocarbons pollute the air every year. Transportation and industrial processes share about equally as main sources of this pollutant.[5]

(3) Nitrogen oxide emissions total about 19.3 million metric tons annually. Over 65 percent of these pollutants come from vehicles and from large coal-fired facilities.[5]

Natural gas can play a significant role in reducing these ozone-causing pollutants. The conversion of vehicles powered by gasoline to compressed natural gas (CNG) can reduce carbon monoxide emissions by 82 percent, reactive hydrocarbons by up to 87 percent, and nitrogen oxides by about one-third. Advanced dedicated natural gas engines are likely to show further improvement in reducing emission of pollutants.

Trucks and buses emit even larger amounts of ozone-causing air pollutants than passenger vehicles. Their conversion to compressed natural gas would be particularly beneficial to the environment.[9] The Brooklyn Union Gas Company has made two buses equipped with dedicated natural gas engines available for service in New York City. These buses are being field-tested by the Command Bus Company in an arrangement with the New York City Department of Transportation. Initial results indicate that these buses are far better for the environment than their diesel-fueled counterparts.[10]

In stationary applications, natural gas can reduce ozone-forming pollutants in a cost-effective manner. New technologies for maximizing energy efficiency, such as natural gas fired cogeneration and combined cycle electric power plants, provide additional environmental benefits by minimizing the amount of fuel required per unit of useful energy. The select use of natural gas at coal-burning facilities can contribute significantly to lower ozone formation at less cost than competing technologies. This matter will be discussed more fully in the next section.

The reduction of ozone at ground level has been given top priority by the federal government as well as by many local authorities. Natural gas can help achieve this goal quickly in a relatively inexpensive manner.

Partnership Between Coal and Gas Protects the Environment

Coal is a plentiful fossil fuel in the U.S. However, coal has major disadvantages to the environment. Its burning generates large amounts of such pollutants as sulfur oxides, nitrogen oxides, and particulates. The mixture of these pollutants with moisture in the atmosphere is believed to be a primary cause of acid rain. The select use of natural gas cofired with coal can greatly reduce these harmful emissions.[11]

The addition of relatively small amounts of natural gas to coal improves burning efficiency, which reduces pollution. Tests have shown that cofiring ten percent gas and ninety percent coal can reduce nitrogen oxides by twenty-five percent and sulfur dioxide by twelve percent.[12,13]

A recently developed technology involves the injection of natural gas into the upper elevation of the coal burning facility. This procedure is called "reburning." By using twenty percent natural gas and eighty percent coal, reburning may reduce nitrogen oxide emissions by as much as sixty percent. When combined with the injection of a dry calcium based sorbent, this procedure can also cut sulfur dioxide emissions in half.[12,13]

It is noteworthy that these cofiring techniques reduce oxides of sulfur and nitrogen by substantially larger amounts than can be attributed to the cleanliness of the natural gas alone. While more complete combustion of the coal is part of the reason, it does not provide the complete explanation. Studies have shown that methane, the principal ingredient of natural gas, has such a strong affinity for oxygen that it limits the amount of that molecule available for the production of sulfur oxides and nitrogen oxides.[12]

To sum up, cofiring of small amounts of natural gas with coal reduces air pollution and acid rain formation in three ways: (1) The cleanliness of natural gas itself; (2) More complete combustion of coal; and (3) The strong affinity of natural gas (CH_4) for oxygen, which reduces the ability of sulfur and nitrogen to form oxides. These beneficial environmental effects can be achieved with minimal capital investment at modest expense. The cost of sulfur oxide control using natural gas has been calculated to be as low as $50 per ton of SO_2 removed, with many potential sites in the $200-$400 per ton range. In contrast, the cost of alternative control strategies ranges from $250 per ton of SO_2 to more than $2,000 per ton.[14]

The select use of natural gas with coal has the additional advantage of reducing *all the major pollutants from coal simultaneously*. These pollutants include sulfur oxides, nitrogen oxides, particulates, carbon monoxide, and carbon dioxide. Moreover, natural gas does not produce any solid waste, such as sludge, ash, or residue, which is a problem with some other pollution control devices.[14] Altogether, the select use of natural gas with coal can greatly reduce environmental pollution, including the formation of ozone and acid rain.

The multiple benefits to the environment, combined with the cost advantages, make the select use of natural gas the best option available to many coal burning facilities in the electric power generating and industrial boiler fields.

The overall level of air pollution in a given area depends on the type of fuel burned as well as on seasonal factors. By combining natural gas facilities, which cause minimum air pollution, with coal-burning ones, the general environment can be kept at acceptable levels. The increased use of natural gas during summer months is particularly desirable, because warm temperatures tend to worsen pollution problems, including ozone formation and acid rain deposition. Fortunately, natural gas is in plentiful supply during the summer, when the space heating market is dormant.

Traditionally, coal and natural gas have been competitors for energy markets. However, environmental constraints and economic considerations make it increasingly likely that coal and natural gas will become partners in producing energy that is cost-effective and environmentally desirable.

Gas Reduces Greenhouse Effect

It is widely believed that increased atmospheric concentration of carbon dioxide and other gases traps solar energy, which results in higher average temperatures at ground level. This phenomenon, which has been call the "greenhouse effect," may have adverse consequences on agriculture and on other activities closely affected by weather conditions. While the theory has not been definitively proven, it is receiving increased attention from scientists, the media, and policy makers. Whatever the ultimate determination, the burning of natural gas generates less carbon dioxide and fewer other gases than any other fossil fuel.[15]

The formation of carbon dioxide (CO_2) is largely determined by the ratio of carbon to hydrogen in the fuel being burned. Methane (CH_4), the main ingredient of natural gas, has the lowest carbon/hydrogen ratio of any fuel. In contrast, oil and coal have substantially higher concentrations of carbon in their molecules. Moreover, methane burns more efficiently than either oil or coal. As a result, the burning of methane results in lower carbon dioxide emissions than the other fossil fuels. In boilers, natural gas emits thirty percent less carbon dioxide than oil per unit of energy. In comparison to coal, natural gas generates half as much carbon dioxide on an energy-equivalent basis.[16]

Other gases that may contribute to the greenhouse effect include nitrogen oxides, carbon monoxide, and chloroflurocarbons. In most applications, nitrogen oxide emissions from natural gas are 35-65 percent lower than from other fossil fuels.[17] Carbon monoxide is caused primarily by emissions from vehicles using gasoline or diesel fuel. The U.S. Environmental Protection Agency estimates that vehicles powered with natural gas would cut carbon monoxide emissions in half.[18] A recent study by Princeton University's Department of Mechanical and Aerospace Engineering found that with a proper control system carbon monoxide emissions "can be effectively eliminated from the exhaust of natural gas fueled engines."[18]

Most technologies involving natural gas do not use chloroflurocarbons. As a result, this pollutant does not normally occur when natural gas is burned. This reality is beneficial in connection with safeguarding the ozone layer in the upper atmosphere, which protects the earth from harmful solar radiation.[16]

While the burning of natural gas minimizes the greenhouse effect, *unburned methane* may contribute to the problem. Being lighter than air, unburned methane escapes into the atmosphere, where it may trap solar energy. Most unburned methane comes from such natural sources as termites, ruminant animals (cattle, sheep), coal seams, and from the conversion of organic debris by anaerobic microorganisms into methane near the surface of the earth, including landfills and rice paddies. Some of this methane, particularly the large quantities that are often present in coal seams, can be commercially recovered and added to gas supplies. Similarly, methane from landfills can be used as a commercial source of fuel. Once methane enters the gas distribution system, virtually all of it is burned and the escape of unburned methane into the atmosphere is minimal.[14]

Overall, natural gas makes a significant contribution to reducing the greenhouse effect.

Friends of the Environment Like Natural Gas

The beneficial environmental qualities of natural gas have received increased recognition from environmentalists. For example, Sierra Club Chairman Mike McClosky called natural gas "the most clean-burning of the fossil fuels" which can "play a significant part of an overall emission control strategy for both stationary and mobile sources."[19]

Similar comments were made by Senator George Mitchell, the Senate Majority Leader, who supports legislation favoring a clean environment. He noted that natural gas can contribute significantly to solving air quality problems. He emphasized the value of existing technologies for using natural gas as an environmentally beneficial fuel.[19]

Congressman Henry A. Waxman, Chairman of the Health and Environment Subcommittee of the House Committee on Energy and Commerce, stated that "the use of natural gas and clean air go hand in hand. "[20] He also noted that natural gas should be specifically identified as a pollution control measure "for both stationary and mobile sources."[20]

President Bush has come out strongly in favor of a clean environment. On June 12, 1989, he stated "we will make the 1990's the era for clean air." He called for the most far-reaching changes in environmental legislation since the Clean Air Act of 1970. He proposed cutting sulfur dioxide emissions from coal-burning power plants almost in half. He also recommended cuts in nitrogen oxide pollution.[21] Natural gas could play a key role in this effort, because it is the most cost-effective means for achieving these goals.

To deal with carbon monoxide and ozone pollution, President Bush called for decisive actions to meet reduced levels mandated by previous legislation, including greater use of alternative fuels which burn more cleanly than gasoline or diesel fuel.[21] Compressed natural gas (CNG), a clean-burning vehicular fuel, could be a major beneficiary from this proposal.

In view of its many environmental advantages, it is likely that natural gas will increasingly emerge as the preferred fuel for many applications.

Sources:

[1] Compilation of Air Pollution Emission Factors, *Third edition, Environmental Protection Agency, May 1978.*

[2] Monthly Energy Review, *May 1978, U.S. Department of Energy.*

[3] Environmental Impact, Efficiency, and Cost of Energy Supply and End-Use, *Volume 1, Pittman Associates, Inc. for the National Science Foundation, the Environmental Protection Agency, and the Council on Environmental Quality, Columbia, Maryland, November 1978.*

[4] Review of the National Ambient Air Quality Standard for Ozone, *U.S. Environmental Protection Agency, March 1986.*

[5] National Air Quality and Emissions Trends Report, 1986, *U.S. Environmental Protection Agency, February 1988.*

[6] *"RRS Fact Sheets: Seasonal VOC Control," U.S. Environmental Protection Agency, Regulatory Reform Staff, June 3, 1986.*

[7] *Testimony Before the Subcommittee on Health and Environment, Committee on Energy and Commerce, U.S. House of Representatives, by Thomas, Lee M., Administrator, U.S. Environmental Protection Agency, February 19, 1987.*

[8] An Assessment of the Risks of Stratospheric Modification, *U.S. Environmental Protection Agency, Office of Air and Radiation, January 1987.*

[9] *"Ozone and Carbon Monoxide: The Role of Natural Gas in Attaining Clean Air Act Compliance, 1988 Update." American Gas Association, March 25, 1988.*

[10] *The Brooklyn Union Gas Company, Natural Gas Vehicles Section, 166 Montague Street, Brooklyn, New York 11201.*

[11] *Remarks by Ralbern H. Murray, Vice Chairman, Consolidated Natural Gas Company, participant in panel discussion on Developments in Natural Gas Cofiring Technology, American Gas Association, Annual Meeting, June 6, 1988.*

[12] *"An Update on the Cheswick Gas/Coal Cofiring Experiences," by Steven Winberg, Consolidated Natural Gas and Berhard P. Breen, Energy Systems Associates, in* Proceedings: Efficient Electricity Generation with Natural Gas, *sponsored jointly by the American Gas Association and Edison Electric Institute, November 16-17, 1987.*

[13] *"Gas Technologies for Emissions Reduction and Operational Benefits," by F. Richard Kurzynske, Gas Research Institute, Ibid.*

[14] *"Natural Gas and the Environment," American Gas Association, Issues Brief 1987-15, December 18, 1987.*

[15] *"Natural Gas and Climate Change: The Greenhouse Effect," American Gas Association, May 1989.*

[16] *"Carbon Dioxide Emissions from Fossil Fuel Combustion and from Coal Gasification," American Gas Association, September 2, 1977.*

[17] *"Compilation of Air Pollutant Emissions Factors," U.S. Environmental Protection Agency, Office of Air and Waste Management, August 1977.*

[18] *"Guidance on Estimating Motor Vehicle Emission Reduction from the Use of Alternative Fuels and Fuel Blends," Emission Control Technology Division, Office of Mobile Sources, Office of Air and Radiation, U.S Environmental Protection Agency, January 29, 1989.*

[19] *"Unlikely Alliance Promotes Use of Gas to Clean Up Air,"* The Natural Resource, *American Gas Association, Spring 1988.*

[20] *Remarks of Congressman Henry A. Waxman at the Natural Gas Roundtable, April 4, 1989.*

[21] *"President Urges Steps to Tighten Law on Clean Air,"* New York Times, *June 13, 1989.*

21. A More Secure Energy Future

The following table depicts anticipated U.S. energy consumption for the period 1987-2010. The projections are based on the Total Energy Resource Analysis (TERA) of the American Gas Association.[1]

Projected Energy Consumption, 1987-2010
(Quadrillion Btu)

Year	Natural Gas	Oil	Coal	Nuclear	Hydro	Misc.	Total
1987	17.7	32.9	18.0	4.9	3.1	0.3	76.8
1988	18.5	34.2	18.8	5.7	2.6	0.3	80.1
1989	19.0	34.4	18.2	6.4	2.5	1.0	81.4
1990	19.5	34.1	17.7	6.8	2.8	1.1	82.0
1995	21.3	34.7	19.2	6.8	3.4	1.7	87.0
2000	22.2	35.2	21.2	6.7	3.5	2.4	91.1
2005	22.6	35.5	23.7	6.3	3.6	3.1	94.6
2010	23.1	36.2	26.4	5.7	3.6	3.7	98.5

During the period 1987-2010, total energy consumption is projected to grow by about 22 quadrillion Btu (Quads), or a little more than one percent annually. In the same interval, gross national product is expected to grow by more than two percent a year, while industrial production is estimated to increase by about three percent annually. These data indicate that energy efficiency will continue to show significant improvement, particularly in the industrial sector.

Increased demand for electricity is expectd to lead to greater coal consumption, which will rise by about two percent a year. Demand for natural gas is expectd to increase by about 1.5 percent annually. Oil consumption will be almost flat. There will be some increase in nuclear energy, which will peak in the year 2000.

While overall oil consumption will not change much during this period, U.S. dependence on imported oil will grow significantly as economically viable domestic resources are depleted. The following table shows oil import trends in relation to total oil consumption.

Oil Imports in Relation to Oil Consumption, 1987-2010
(Quadrillion Btu)

Year	Consumption	Domestic Prod.	Imports
1987	32.9	19.9	13.0
1988	33.9	19.5	14.4
1989	34.4	19.0	15.4
1990	34.1	18.4	15.7
1995	34.7	16.3	18.4
2000	35.2	15.0	20.2
2005	35.5	14.1	21.4
2010	36.2	13.5	22.7

In 1987, oil imports accounted for about forty percent of consumption. By the mid-1990's, imports are projected to exceed domestic oil production. In the year 2010, imports are likely to account for almost two-thirds of oil requirements.

U.S. oil production is in a declining trend. The resource base of low-cost domestic oil is being depleted. New oil resources in the U.S. are expensive to develop. To place this matter into perspective, it may cost $15 a barrel or more to find and produce new oil in the U.S. In contrast, the cost of producing oil from foreign sources is generally less than $5 per barrel. Fundamentally, U.S. incremental (new) oil is becoming increasingly noncompetitive from an economic standpoint.

Fortunately, the outlook for domestic natural gas is much

more favorable. The U.S. has a very large natural gas resource base, which can be produced economically. Methane gas is the most substitutable energy form for oil. To assure a more secure energy future, it makes good sense to increase the use of natural gas as a replacement for excessive dependence on imported oil. The following table shows the potential replacement of petroleum products with natural gas, based on studies by the American Gas Association.[2]

Potential Substitution of Oil with Natural Gas
(thousand barrels/day)

Stationary Sources	**Immediate**	**One Year**	**5 Years**
Residential	0	40	250
Commercial	50	100	310
Industrial	50	140	285
Electricity generation	60	200	460
Subtotal	160	480	1,305
Vehicles	0	0	30
Total	160	480	1,335

The preceding table shows that natural gas could quickly replace 1.3 million barrels of imported oil a day in stationary applications. However, under present conditions very little could be done to help the transportation sector, which depends almost exclusively on fuels derived from oil, notably gasoline and diesel fuel.

Compressed natural gas (CNG) is an excellent vehicular fuel, but its increased use requires an infrastructure of filling stations which is not yet in place. Moreoever, vehicles have to be adapted to use natural gas, and/or manufacturers have to produce vehicles designed with natural gas engines.

It is in the national interest, both from the standpoint of energy security and in terms of environmental benefits, to implement a program that will create the infrastructure for natural gas vehicles as soon as possible. When such a policy becomes a reality, natural gas can play a major role in replacing imported oil in the transportation sector. It would be desirable to set a goal of using natural gas as the primary fuel for ten percent of the vehicular market by the end of the twentieth century.

If such policies are implemented, the use of natural gas for vehicles could displace 700 thousand barrels of oil/day. When combined with increased use of gas in stationary applications, total gas demand could rise by 4 Tcf, which is the equivalent of 2 million barrels of oil/day. This development would create the conditions for energy leadership by natural gas some time in the early part of the twenty-first century.

Sources:

[1] *"The A.G.A. TERA 1989 Mid-Year Base Case," American Gas Association, August 16, 1989 (draft).*

[2] *"The Role of Natural Gas in Offsetting Oil,"* Energy Analysis, *American Gas Association, October 12, 1988.*

22. Gas Price Trends Favor Increased Demand and Supply

In a free market environment, commodity prices are determined by the interplay between demand and supply. Under normal circumstances, prices go up when demand is strong; they will go down when demand weakens. Interference with the pricing mechanism by the government tends to produce distortions which lead to undesirable consequences. For example, between 1954 and 1973 the Federal Power Commission kept natural gas prices at the wellhead so low that producers lacked the incentive to drill for sufficient reserves to meet requirements. As a result, gas shortages occurred in the mid-1970's. In contrast, the gas price escalation that took place in the early 1980's swung the pendulum in the other direction. Demand for gas declined significantly, particularly in the industrial sector, while the supply rose rapidly. This combination resulted in excess supplies in the form of a so-called "gas bubble," which has required several years of reduced prices to find markets. Fortunately, the trend toward a free market in gas prices is firmly in place. According to legislation passed by the Congress and signed by the President, the remaining price controls on natural gas are to be removed by January 1, 1993.[1]

During the period 1987-2010, natural gas consumption is expected to increase from 17.7 Quads to 23.1 Quads, which averages out to about 1.2 percent per annum. This increased demand can be met with field price increases averaging about six percent annually (in constant dollars) between 1990 and 2010. In the same interval, prices to residential and commercial gas customers are estimated to rise only about three percent a year. This modest increase in consumer prices will be made possible by anticipated efficiency improvements in utilizing the existing natural gas transmission and distribution system. Increased markets for year-round applications like water heating and cogeneration, as well as greater penetration of the air conditioning markets in the summer months, will help balance the gas load and make possible savings that can be passed on to consumers. The cost of gas to industrial users, who have benefited from deeply discounted prices during the "gas bubble" era, is expected to rise about three percent annually. A similar price pattern is likely to prevail for the use of gas in connection with the production of electricity. The following table presents details on anticipated price trends.[2]

Estimated Natural Gas Prices, 1987-2010
(constant value 1989 $/MMBtu)*

		Delivered Prices				
Year	**Field Price**	**Residential**	**Commercial**	**Industrial**	**Electric Utilities**	**Average**
1987	1.62	5.37	4.64	2.85	2.25	3.93
1988	1.66	5.32	4.50	2.87	2.27	3.97
1989	1.61	5.59	4.68	2.88	2.21	3.74
1990	1.81	5.66	4.77	3.01	2.31	3.82
1995	2.18	6.05	5.16	3.45	2.69	4.20
2000	2.87	6.79	5.89	4.22	3.43	4.95
2005	3.77	7.77	6.87	5.24	4.44	5.94
2010	4.98	8.95	8.04	6.45	5.63	7.13

***dollars per million British thermal units**

The above data are average figures for the nation as a whole. The delivered prices will vary with the distance the gas has to be transported from the point of origin to the consuming market.

It is noteworthy that much of the anticipated consumer price increases in natural gas can be offset through improved energy efficiency. By switching to more fuel-efficient equipment, the average consumer can keep monthly gas bills essentially flat during the next two decades. (These comments apply to constant 1989 dollars; they do *not* include inflation adjustments.) Moreover, consumers can increase their fuel savings by switching from electricity to gas for such applications as space heating, hot water production, air conditioning, and cooking. This analysis shows the great importance of improved energy efficiency as the key for keeping expenses under control.

The increased field prices of natural gas are expected to be necessary to motivate producers to develop higher cost resources, including nonconventional sources of methane.

Fair prices to producers and consumers are the best foundation for assuring adequate supplies to meet the growing demand for natural gas.

Sources:

1 *"House Passes Bill to Decontrol Prices for Gas,"* Wall Street Journal, *April 18, 1989.*

2 *"Historical data from "Monthly Energy Review: April 1989;" Projections from "The A.G.A. TERA Mid-Year Base Case," American Gas Association, August 16, 1989 (draft).*

23. Gas and Electricity

Gas and electricity are interrelated in complex ways. They are major competitors in most energy markets. Gas plays an important role in producing electricity. Many electric utilities are also in the gas business. Gas companies are increasingly involved with cogeneration, an efficient technology for producing electricity. Gas is considerably cheaper than electricity in most applications. Both electric and gas utilities would benefit from greater use of gas and less demand for electricity during summer months, when electric generating facilities are stretched to capacity by air conditioning requirements.

For the past century, gas and electricity have been competitors in many energy markets, including residential, commercial, and industrial applications. Both sources of energy are convenient to use. The main differences are in their relative prices and in the range of products (appliances) available to consumers. Gas is considerably cheaper than electricity, particularly under present and foreseeable conditions. However, electric appliances have had the advantage of greater variety and stronger support from leading manufacturers, who are also major suppliers of electric generating equipment. If gas is to increase its market share and to capture new markets, it is imperative to offer consumers additional appliances and equipment.

The generation of electricity is a very capital-intensive enterprise. It costs about 2 to 4 times as much capital to produce and deliver electric energy to customers as it does to offer them the energy equivalent in the form of gas.[1] Moreover, the traditional production of electricity wastes considerable amounts of energy. About two-thirds of the fossil fuel (coal, oil, or natural gas) used in the production of electricity is lost in the form of waste heat. It is apparent that the economics of electricity production is heavily influenced by the cost of capital and by the availability of cheap energy sources.

The oil price escalations of the 1970's have seriously undermined the economic foundations of conventional electricity generation in giant, centralized facilities. The era of cheap energy has come to an end and is unlikely to recur any time soon, if ever. All the fuels used in the production of electricity have gone up considerably in price. These energy price increases have had a negative impact on the electric utility industry and its customers, who pay for all the energy that is used, including the amounts that are wasted in the process.

The energy price escalation was accompanied by sharp increases in interest rates, which had adverse effects on capital costs. Whereas prior to 1973 electric utilities could borrow long-term money at about 5 percent, since that time interest on such borrowings have averaged closer to 10 percent. The problem has been compounded by the escalation of construction costs.

Higher costs of energy and of borrowed funds have placed electricity at a disadvantage in relation to natural gas. The following tables compare costs of gas and electricity in major markets.

Residential Market[1]
(1989 $/MMBtu)

Year	Natural Gas	Electricity
1987	5.80	23.44
1988	5.56	22.96
1989	5.59	21.47
1990	5.66	21.73
1995	6.05	23.14
2000	6.79	23.89
2005	7.77	25.09
2010	8.95	26.23

Commercial Markets[2]
(1989 $/MMBtu)

Year	Natural Gas	Electricity
1987	5.01	22.17
1988	4.71	21.49
1989	4.68	20.54
1990	4.77	20.68
1995	5.16	21.48
2000	5.89	22.15
2005	6.87	23.21
2010	8.04	24.22

Industrial Markets[2]
(1989 $/MMBtu)

Year	Natural Gas	Electricity
1987	3.08	14.93
1988	3.00	14.16
1989	2.88	14.10
1990	3.01	14.12
1995	3.45	14.64
2000	4.22	15.09
2005	5.24	15.78
2010	6.45	16.44

It is apparent from these tables that electricity prices are two-and-a-half to four times greater than gas prices to residential, commercial, and industrial customers. Moreover, very little improvement in that ratio is anticipated during the next two decades. Under these conditions, it is likely that consumers will turn increasingly to the cheaper fuel, particularly as new energy-efficient gas appliances and equipment will be offered in the marketplace.

Sources:

[1] *"Repowering with Natural Gas—An Electricity Generation Option,"* Energy Analysis, *American Gas Association, Oct. 17, 1986.*

[2] *Historical data from* Monthly Energy Review, *April 1989; projections from the "A.G.A. TERA 1989 Mid-Year Base Case," August 16, 1989 (draft).*

24. Economic Advantages of Gas Over Oil

From the wellhead to the point of end use, natural gas enjoys significant economic advantages over oil. U.S. natural gas resources are more plentiful than those of oil; in most cases, it is more economical to drill for gas than for oil. The primary recovery rate for natural gas wells is substantially higher than that for oil wells. Natural gas can be utilized essentially the way it comes out of the ground, while oil has to be refined. The natural gas delivery system makes this fuel available to customers at all times, while oil has to be ordered periodically. For the consumer, natural gas involves no inventory costs, storage, or advance payments, while oil does. The removal of pollutants *before* gas enters the pipeline makes it the cleanest burning fuel, which reduces cleaning and equipment maintenance costs in comparison with oil. Because of these economic advantages, natural gas is the preferred fuel for most stationary applications. Once the infrastructure is in place, natural gas can also compete successfully with oil-derived fuels for transportation markets.

The large natural gas resource base of the U.S. makes drilling for gas more productive than drilling for oil. In recent years, newly drilled gas wells have yielded three times as much energy as new oil wells.[1]

Primary recovery from gas wells averages seventy percent of the gas in place, compared with twenty-five percent for oil wells.[2] Natural gas, being lighter than air, comes to the surface on its own. In contrast, oil requires pressure from other sources to move to the surface. In the primary recovery phase, oil is pushed to the surface largely by the pressure from natural gas that is associated with oil in the same field. To increase production from oil wells, a variety of measures are employed, including injection of water and/or gas (secondary recovery), and the use of chemicals or heat (enhanced oil recovery). The employment of these techniques may bring oil recovery up to fifty percent of the oil in place, still substantially less than primary recovery from gas wells. All efforts to increase oil output through external stimulation add to the cost of production.

One of the unique features of natural gas is its readiness for end-use applications without much further processing. In effect, natural gas has been refined to a considerable extent by nature. In contrast, crude oil is generally not used in its original state. It is usually processed in refineries and transformed into such products as gasoline and diesel fuel before it is sold to customers. These refining procedures add to the cost of the end products. It follows that realistic cost comparisons of gas should be made with refined products, not with crude oil.

In comparison with oil, gas is more convenient and saves inventory expense. Households connected to natural gas have the fuel available at all times. They do not need to order it, store it, or pay for it in advance. Oil users have to go through these procedures to have the fuel when they need it.

The structure of methane, consisting of one carbon atom and four hydrogen atoms, gives it significant heating advantages over oil. According to a report by the Department of Energy/Federal Energy Regulatory Commission, the total cycle efficiency for space heating ranges from 44-53 percent for methane, compared with 33-39 percent for fuel oil.[3] It takes about four units of oil to generate the same amount of heat as three units of methane.

Before natural gas is allowed to enter the pipelines, undesirable pollutants such as sulfur are removed. As a result, natural gas burns more cleanly than oil products. In addition, gas equipment remains at optimum levels of efficiency for longer periods of time than oil equipment. The latter has to be frequently cleaned and serviced to avoid loss of efficiency.

These economic advantages of natural gas make it the preferred fuel for most stationary applications. Even in the transportation sector, which is currently monopolized by oil-derived fuels, natural gas can compete effectively once the infrastructure of filling stations and vehicles with dedicated gas engines is in place.

For more than a century, oil has dominated the energy economy, while natural gas was treated as a stepchild. This economic dominance by oil has also affected the attitudes of the general public and the views of opinion leaders. The erroneous impression has been created that oil is a better fuel than natural gas. In actuality, as has been shown in this chapter, in most respects natural gas has significant economic and other advantages over oil. A better understanding of the truth about natural gas will help this fuel achieve energy leadership in the future.

Sources:

[1] U.S. Crude Oil, Natural Gas, and Natural Gas Liquids Resources, *Energy Information Administration, 1987.*

[2] *"The Future of Deep Conventional Gas," by John M. Hunt, UNITAR Conference on Long-Term Energy Sources, Montreal, December 1979, Pitman Publishing Company, Marshfield, MA 02050, published in 1981.*

[3] Natural Gas Survey: Efficiency in the Use of Gas, *Department of Energy/Federal Regulatory Commission, June 1973.*

25. The International Gas Boom

Between 1970 and 1987, world gas consumption outside the United States more than tripled, from 14.2 to 45.4 Quads.[1] The oil price escalation of the 1970's provided a major impetus for this upsurge in gas consumption. Other factors which played a role in the popularity of natural gas include the versatility of this fuel, its availability from many sources of supply, and its environmental advantages. Natural gas has the best public image of any fossil fuel.

The following table lists nations that consumed more than 1 Quad of natural gas in 1987.

Consumption of Natural Gas, 1987[1]
(in trillion Btu)

USSR	21.0
USA	17.5
Western Europe	8.4
United Kingdom	2.0
West Germany	1.8
Canada	1.7
Japan	1.5
Netherlands	1.4
Italy	1.3
France	1.0

The Soviet Union, which consumed only 6.2 Quads of natural gas in 1970, has jumped to the lead with 21.0 Quads in 1987. In contrast, the USA, which was at the top of the list in 1970 with gas consumption of 22 Quads, has fallen to second place in 1987. The USA was the only major country which showed a decline in gas consumption between 1970 and 1987.

The Soviet Union has also taken a commanding lead in gas production. In 1987, the USSR produced 26.4 Quads of natural gas, compared with 16.8 Quads by the USA. The extraordinary rise in gas production, consumption, and exports by the Soviet Union reflected major policy decisions by the Soviet government, which has given top priority to natural gas.[2]

The following table cites the world's leading gas producers.

Production of Natural Gas, 1987[1]
(in trillion Btu)

USSR	26.4
USA	16.8
Canada	2.7
Netherlands	2.3
United Kingdom	1.6
Algeria	1.5
Indonesia	1.2
Norway	1.1
Mexico	1.0

Other major producers include Australia/New Zealand (0.7), Venezuela (0.7), West Germany and Italy (0.6 each), and China (0.5). It is noteworthy that several countries in Western Europe (Netherlands, United Kingdom, and Norway) are among the major producers of natural gas. Moreover, West Germany and Italy produce a significant amount of their own gas requirements. Natural gas provides a much higher degree of energy security to the industrial nations than oil.

Major natural gas importers include the industrial nations and countries in Eastern Europe. The following table provides details.

Importers of Natural Gas, 1987[3]
(in trillion Btu)

West Germany	1.6
Japan	1.4
France	1.0
USA	0.9
Italy	0.8
United Kingdom	0.4
Czechoslovakia	0.4
Belgium, Luxemburg	0.3
Poland	0.3
East Germany	0.3
Bulgaria	0.2
Yugoslavia	0.2
Hungary	0.2

In 1987, 6.8 Quads of natural gas were exported by pipeline. The Soviet Union accounted for almost half the total. The following table presents additional information.

Natural Gas Exports by Pipeline, 1987[3]
(in trillion Btu)

USSR	3.0
Netherlands	1.3
Norway	1.1
Canada	0.9
Algeria	0.4

A significant trade in liquefied natural gas (LNG) has also developed during the past 17 years. Such shipments grew from 0.1 Quads in 1970 to almost 2 Quads in 1987. The following nations accounted for most of the LNG shipments.

Liquefied Natural Gas Exports, 1987[3]
(in trillion Btu)

Indonesia	0.763
Algeria	0.482
Malaysia	0.292
Brunei	0.256
Abu Dhabi	0.104

Japan obtains most of its natural gas in the form of LNG. It is too far from sources of supply to make pipeline construction economical.

Proved resources of major natural gas producing countries total about 3,000 Quads, enough to supply almost fifty years gas requirements at current levels of consumption. The total conventional gas resources of the world are much larger than the proved reserves. For example, in the U.S. estimated conventional gas resources are about five times the proved reserves. In other parts of the world, where much less is known about natural gas resources than in the U.S., the ratio is likely to be even higher. On the basis of these data, it seems reasonable to assume that conventional natural gas resources will be sufficient to take care of global requirements for several centuries. The following table shows the proved natural gas reserves of leading countries.

Proved Gas Reserves in 1987[1]
(in quadrillion Btu)

USSR	1,480
Iran	500
USA	191
Abu Dhabi	187
Qatar	158
Saudi Arabia	148
Norway	108
Canada	83
Netherlands	65
Kuwait	36
China	32
United Kingdom	22

The next table shows proved gas reserves on a regional basis.

Regional Gas Reserves in 1987[1]
(in quadrillion Btu)

Africa	202
Asia	205
Australasia	22
Eastern Europe	1,541
Latin America	234
Middle East	1,105
North America	292
Western Europe	223
Total	3,824

On a global basis, most of the proved gas reserves are situtated in Eastern Europe (primarily the Soviet Union) and the Middle East. However, the other regions of the world are also well endowed with natural gas. Technological developments are likely to increase proved reserves of gas significantly. Hitherto, the United States had led the world in the invention of new approaches and technologies to improved gas recovery and enlargement of the proved reserve base.

In addition to conventional gas resources, the world has even larger amounts of unconventional sources of methane. Technological developments for the recovery of such resources, which are particularly advanced in the USA, will expand the methane gas resource base even further. If mankind acts rationally in relation to energy, adequate amounts of methane gas will be available for the rest of time.

The worldwide increase in the utilization of natural gas has favorable implications for the United Staes. The cleaner environment which follows in the wake of increased gas usage will benefit all parts of the globe. The presence of a large natural gas resource base, and its widespread distribution, will make future energy disruptions less likely.

Gas appliances and equipment are now being manufactured in many parts of the world. Some of these products will benefit gas consumers in the U.S. For example, Japan has developed gas heat pumps and cooling equipment that help meet U.S. needs. Many countries, including Italy, New Zealand, and Canada, have moved heavily into natural gas vehicles, which will facilitate U.S. efforts in that direction. In return, the U.S. can offer the rest of the world high-efficiency gas furnaces and the most advanced cogeneration equipment. U.S. petroleum producers have unexcelled expertise in finding and developing gas resources. Increased international trade in natural gas equipment, technologies, and services should be beneficial to all parties involved.

It seems likely that the worldwide boom in natural gas will continue for many years to come.

Sources:

[1] BP Review of World Gas, *July 1988.*

[2] *"The Soviet Gas Campaign," by T. Gustafson, Rand Corporation, Santa Monica, California, June 1983.*

[3] Annual Bulletin of General Energy Statistics for Europe, *Economic Commission for Europe, United Nations, 1988.*

26. Trends Favoring Natural Gas

The following trends have contributed to making natural gas the best energy choice now and for a long time to come:

(1) Favorable outlook for gas supply.
(2) Reasonable price prospects.
(3) Environmental advantages.
(4) New technologies and applications.
(5) Comparative advantages over competing fuels.
(6) The worldwide growth in gas use and production.

The United States has plentiful conventional gas resources, which should take care of anticipated requirements well into the next century. In addition, advances in geological science and the development of technologies for producing gas from deep locations, from offshore below thousands of feet of water, and from tight formations will greatly enlarge the usable resource base. Similar comments are applicable to the gasification of coal, peat, and oil shale, which can add sufficient methane gas supplies to last for centuries. The renewable sources of methane, including landfills, animal wastes, sewage, and energy crops such as ocean plants, will be increasingly utilized and will take care of U.S. gas energy requirements as long as the sun shines and mankind inhabits the earth.

The outlook for gas prices is favorable in the foreseeable future. Government controls over wellhead prices have already been largely removed; the remainder will be gone by 1993. A free market will determine prices on the basis of demand and supply. The anticipated increase in demand will gradually strengthen gas prices. This development will bring new sources of gas to the market, which will keep prices from getting out of hand.

Generally speaking, unconventional sources of methane gas are more expensive to produce than conventional ones. However, scientific and technological breakthroughs can enlarge the conventionally priced gas resource base by significant amounts. Research and development should play an important role in assuring the availability of adequate gas resources at reasonable prices. The ultimate ceiling on gas prices will be the cost of plentiful methane gas from renewable sources.

Environmental issues have moved to the forefront of concern in the United States as well as all over the globe. The earth is increasingly viewed as an ecological unity that must be safeguarded from harmful pollutants. Methane gas is the cleanest of all fossil fuels. In fact, it burns so cleanly and minimizes pollution formation by other fuels that it has been recognized as a control technology. The cofiring of small amounts of natural gas with coal can greatly improve combustion and pollution reduction, thereby helping coal meet environmental standards.

Because potential pollutants, such as sulfur, are removed from natural gas *before* it enters a pipeline, the expense of installing costly pollution control equipment at the point of end use can be avoided. The environmental advantages of natural gas have favorable economic implications. Pollution will be increasingly recognized as an economic cost. Therefore, a fuel which minimizes pollution has a competitive advantage. Natural gas will be in that fortunate position.

In the past, the petroleum business has been dominated by oil, while natural gas was treated as a stepchild. This orientation defined the role of gas as an adjunct to oil production (to help lift oil to the surface), and as a fuel to be used solely in stationary applications, primarily as a source of heat. Gas was excluded from the transportation market, which was monopolized by gasoline and diesel fuel derived from oil.

In actuality, methane gas has significant advantages over its oil-derived couterparts for transportation applications. Methane has an octane rating of 130, compared with less than 100 for gasoline. Methane burns far more cleanly than gasoline or diesel fuel. From an environmental standpoint, methane gas is the fuel of choice. Moreover, engines fueled with methane have a longer life than gasoline engines. One can make a convincing case that methane is a better transportation fuel than gasoline or diesel fuel. Its main disadvantage is the absence of an infrastructure of filling stations. Once that problem has been solved, people may wonder why methane gas has not been used all along as the preferred fuel.

In stationary applications, electricity has been the main competitor of natural gas. Gas and electricity are equally convenient; they can be turned on and off instantaneously. The main advantages of gas are its comparatively low cost and its environmental benefits. Electricity has benefited from strong support by manufacturers of appliances and equipment. Recent developments increasingly favor gas.

For the past century, electricity has benefited from a stream of new products, appliances, and equipment that broadened its applications. Electricity has been used for lighting, heating, cooling, cooking, drying, and energizing all types of devices. While electricity is uniquely well suited for most lighting applications and for energizing such devices as radios, television sets, and computers, it has no real advantages over gas for heating, cooling, cooking, or drying. The inroad of electricity in those fields has been due primarily to innovative product development and marketing, not any inherent advantages. Consumers could save plenty of money if they switched from electricity to gas for these applications.

The economics of electricity production are largely determined by the cost of fuel and capital. Prior to 1973, both of these costs were low, which favored electricity. However, since the oil and construction price escalations of the 1970's, those costs have turned against electricity. Fossil fuel costs have increased substantially, and the cost of capital (interest charges) has more than doubled. Currently, electricity costs three or four times as much as gas per unit of energy. These economic realities give natural gas a significant advantage.

Manufacturers of applicances and equipment are aware of the fact that gas makes good economic sense and that consumers would benefit from its greater utilization. An increasing stream of innovative, efficient gas appliances is coming to market. The speed with which such developments can occur is demonstrated by high-efficiency gas furnaces and boilers. In 1980, when I wrote my first book on natural gas, there was only one gas furnace in development which had an anticipated efficiency in excess of 90 percent, namely the Lennox pulse combustion furnace. This revolutionary piece of equipment went into production in the early 1980's. Over the next several years, most manufacturers of gas heating equipment developed furnaces with efficiency ratings of 90 percent or more.

Similar developments may be in the offing in relation to gas cooling equipment. Some of the strongest manufacturing enterprises in the country are engaged in research and development of gas heat pumps and air cooling equipment for commercial markets. Natural gas wil increasingly benefit from the involvement of leading manufacturers in the production of innovative, highly efficient equipment that will lead to major inroads in the marketplace.

The worldwide growth of natural gas in the past two decades has been phenomenal. Natural gas has become a leading fuel in virtually all industrial countries as well as throughout Eastern Europe. Many new products have been developed for the efficient use of gas in both stationary and mobile applications. For example, Japanese manufacturers have produced gas heat pumps for a number of years. Similarly, Italian manufacturers have developed advanced technologies for gas-fueled vehicles. The marketplace for gas appliances and equipment will be increasingly international. U.S. consumers will benefit from gas equipment developed in foreign countries, while overseas customers will be able to purchase highly efficient gas furnaces and other equipment offered by U.S. manufacturers. This exchange of gas products across national borders will benefit the natural gas industry everywhere.

The rapid growth of gas in Europe and Japan during the past two decades was greatly facilitated by disenchantment with oil, which had been the world's energy leader. The price escalations and supply disruptions that took place in world oil markets during the 1970's have given oil a bad reputation. Millions of consumers experienced first-hand the adverse effects of these developments on their lives. Moreover, nations heavily dependent on oil imports became concerned about their energy security. They turned to natural gas as a preferable option.

Taking all factors into consideration, natural gas enjoys the most favorable public image of any major energy source. Its environmental credentials are first-rate. Its consumer-friendly features, including ease of use, ready availability, and wide range of applications, add to its positive image. Its comparatively low cost, particularly in relation to electricity, will make it increasingly popular with consumers.

Like Cinderella, natural gas is emerging from obscurity and neglect to a status appropriate to its many positive features and advantages. The time has come to give recognition to natural gas as the best energy choice for many applications.

27. Gas Ranks First

A comparison of various energy sources involves many variables. While objective criteria should predominate, judgmental and other subjective factors cannot be eliminated. Therefore, it is important to clarify the basis for one's views. This procedure will enable the reader to judge the fairness and validity of the approach and/or to make his or her own evaluation.

In the following presentation, natural gas, oil, coal, and electricity will be evaluated in terms of the criteria cited below.

(1) Production costs.
(2) Transportation costs.
(3) Environmental costs.
(4) Consumer values.
(5) End use convenience.
(6) Appliances and equipment.
(7) Versatility.
(8) Energy security.
(9) Public image.

Each factor will be ranked in a range from 1 to 10. The more desirable the rating, the higher the number. For example, the lowest environmental costs will be given a rating of 10; the highest environmental costs will be assigned a 1. The process of examining and assigning values to these variables will be helpful in gaining a better understanding of these energy sources, even if there may be disagreement on some ratings.

Production Costs

The cost of finding gas ranges from modest to very high, depending on where the exploration takes place. In a shallow field with proven resources, costs are relatively low. In contrast, exploring for deep gas (below 15,000 feet) in an unproven area can be very expensive. Once gas has been found, it is relatively easy to recover, as it comes to the surface on its own. Primary recovery averages 70-80 percent of the gas in place. Processing costs tend to be modest. In the case of sour gas, sulfur has to be separated from the gas. While this procedure involves some cost, it also yields revenues from the sale of high purity sulfur, which is a desirable industrial product. As a general operating procedure, impurities are removed from gas before it is allowed to enter the pipeline. As a result, gas is very clean and environmentally benign when it reaches consumers. Overall, gas production costs are modest and deserve a rating of 8.

The U.S. oil reserve base has been more thoroughly explored than that for natural gas. Therefore, it is more costly to find economically viable sources of new oil than of gas. Moreover, primary recovery of oil is much lower than that of gas, averaging 15-25 percent (compared with 70-80 percent for gas). To recover additional amounts of oil, costly procedures are necessary. Oil has to be refined into usable products before it is ready for the market. This procedure adds substantially to the cost. Overall, oil production costs in the U.S. are fairly high and are assigned a rating of 5.

The location of coal seams is well known; there is virtually no risk of a "dry hole" in coal production. The coal extraction process is highly mechanized, but it involves much more labor than either gas or oil production. Labor is particularly intensive in connection with deep coal mining. The preparation of coal for end use involves breaking the material into small pieces and washing them. Coal production disturbs larger areas of the environment than gas and oil production. Coal production costs are moderately high. The rating is 5.

The production of electricity involves the conversion of fuels or gravity into usable energy. Major sources of electricity include coal, nuclear energy, and hydroelectric facilities. All of these require very heavy capital investments. It generally takes several years to obtain the necessary permits for building a new electric power station. In connection with coal, considerable land is required for storage and handling, as well as for disposal of ash and sludge. Pollution control equipment is mandatory on all new facilities. The use of coal also involves large amounts of water. High costs of capital have hampered the construction of new coal-burning facilities in recent years. The costs and risks involved in nuclear power plants are so high that no new orders for such facilities have been placed since 1978. The availability of hydroelectric facilities is limited by geographical factors. It involves the building of dams, which is costly and often involves environmental problems. High capital expenses for traditional baseload power plants, whether coal, nuclear, or hydro, give electricity a low rating of 2. New technologies, such as combined cycle power production and cogeneration, using highly efficient gas equipment, reduce capital and operating costs substantially. Once their use becomes more widespread, the production cost rating of electricity will rise to a higher level.

Transportation Costs

Natural gas is transported from producing fields to consumers via 1,150,000 miles of pipelines and mains. While this transmission system is highly efficient, it involves significant costs, which include not only transportation service, but also storage and load balancing. For example, in 1987 the wellhead price of gas was $1.71 per thousand cubic feet (Mcf); the average price at the city gate was $3.34 per MMBtu.[1] Transportation costs vary with the distance from the wellhead. For residents of Texas, Oklahoma, and Louisiana, where most of the gas originates, transportation costs are so low that a rating of 10 would be justified. In contrast, residents of the

Northeast would give a low rating to transportation costs. Overall, a rating of 6 seems appropriate.

Crude oil has to be shipped to refineries for processing. From there, the products are forwarded to distribution facilities, which supply consumers. About two-thirds of oil products are used in the transportation sector (gasoline, diesel fuel, and aviation fuel). Oil products are transported by ship, barge, railroad tank cars, and trucks. The transportation system for oil is efficient and less expensive than for gas. It is given a rating of 8.

Coal is generally used by electric utilities or by large industrial power plants. These users tend to be located near the coal mines or in places that have access to low-cost water transportation. Most coal is transported by special trains, which shuttle between the mines and the user. While these procedures reduce expenses, transportation costs are still high because of the bulky nature of coal. A rating of 6 seems appropriate.

Electricity is transmitted from generating stations to consumers via cables and wires. The cost of constructing these transmission facilities is high in capital, but relatively low in operating expenses. About nine percent of the electric power is lost in the transmission process. Eight seems to be a suitable rating.

Environmental Costs

Because natural gas is the cleanest fossil fuel, its use involves the lowest environmental costs. Gas requires no costly pollution control devices. Moreover, gas equipment tends to last longer and needs less cleaning and maintenance than that used with other fuels. An environmental rating of 10 is well deserved by gas.

Oil products have significant environmental problems. Gasoline and diesel fuel are major sources of carbon monoxide, nitrogen oxides, reactive hydrocarbons, and particulates. These fuels play a major role in causing ground-level ozone (smog), which is a significant hazard in many urban areas. In stationary applications, No. 2 fuel oil, which is used in residential and commercial furnaces and boilers, emits more pollutants than gas and necessitates more frequent cleaning of the equipment. The burning of residual fuel oil, which is mostly used in industrial applications, involves considerable emission of pollutants. The environmental rating of oil products is a relatively low 4.

When coal is burned, it emits large quantities of pollutants, including sulfur oxides, nitrogen oxides, and particulates. The large-scale use of coal for generating electricity is considered a primary cause of acid rain and an important contributor to the greenhouse effect. To reduce harmful emissions, pollution control equipment is frequently used. This procedure adds to the cost of using coal. Coal also has the problem of generating large quantities of sludge and ash, which must be removed. Waste disposal procedures add to operating costs. From an environmental standpoint, coal has the lowest rating of 1.

Electric utilities are the largest users of coal. Therefore, they inherit the problems cited in the previous paragraph. Because of increasing public concern about pollution, environmental costs of utilities are going to rise in the future. In connection with the clean air legislation recently introduced by the Bush Administration, the Environmental Protection Agency estimates the added cost to the electric utility industry for reducing pollution would be $1 billion a year starting in 1996 and $4 billion a year after 2003.[2] In addition to air pollution from coal, electric utilities have to deal with radioactive waste from nuclear facilities, which also poses serious environmental problems. In terms of the environmental costs involved with electricity production, the lowest rating of 1 would be appropriate. However, this negative view is mitigated by the fact that at the point of end use, electricity is very clean. We will give it an overall rating of 3.

Values to Consumers

It is important to differentiate between value and price. Value is a function of price plus all the other factors that affect the total cost of energy to consumers.

High efficiency, environmental cleanliness, convenience, and other user-friendly features, combined with reasonable price, make natural gas the best energy value for many applications. Competing energy sources either lack these positive features or have a much higher price. Consumers can obtain better energy values and achieve substantial savings by switching from oil or electricity to natural gas for space heating, cooling, hot water production, cooking, and drying. Gas is rated 10 for value.

While the price of No. 2 fuel oil is similar to that of natural gas, oil has a lower value than gas. Because it generates soot, oil causes equipment to lose efficiency and to require frequent cleaning. Oil must be ordered and paid for in advance and stored at the user's premises. It is less clean, less convenient, and less user-friendly than gas in stationary applications. In the transportation sector, prices of gasoline and diesel fuel are higher than the cost of compressed natural gas. Gasoline and diesel fuel are major polluters, which lowers their value. Oil is given a value rating of 5.

Coal has the lowest price per unit of energy, but its value to consumers is reduced by high costs of capital, handling charges, and pollution control expenses. On the positive side, coal's price stability, usually assured by long-term contracts, adds to its value. An overall value rating of 7 seems appropriate.

At the point of end use, electricity is efficient, clean, and user-friendly. However, its cost is three to four times higher than that of gas, which has similar qualitative features. In comparison with gas, electricity is not a good value. Wherever these energy sources compete directly, consumers could generally improve energy values by switching from electricity to gas. In relation to gas, electricity would get a low rating of 3 or 4. However, electricity deserves the highest rating for lighting and for powering television, computers, and many other devices. A rating of 7 takes into account its positive features and its high price.

End Use Convenience

Gas is always available to the consumer at the point of end use. It does not need to be ordered, nor is it payable until after it has been used. There are no waste products which need to be removed. Gas is very consumer friendly and deserves the highest rating of 10.

Oil lacks many of the advantages cited for gas. Fuel oil has to be ordered and paid for in advance of use. It has to be stored at the place of consumption. The storage tank has to be cleaned periodically. It also has to be kept leak-proof to avoid environmental contamination. In the transportation sector, gasoline and diesel fuel are sold through filling stations that cover the country along highways and in population centers. These filling stations do an outstanding distribution job. They qualify for a rating of 10. However, in stationary applications, a much lower rating would be appropriate. Overall, for end use convenience oil is rated 7.

The use of coal involves many inconveniences, including cleaning and preparation before burning, pollution control during combustion, and the removal of ash and sludge afterwards. All of these procedures are not only inconvenient, but also costly. End use convenience of coal ranks at the bottom with a rating of 1.

Electricity gets a top rating for convenience. By turning a switch, the electric current goes on or off. One potential problem that could reduce the top convenience rating involves brown-outs, which result from equipment breakdown and/or excessive demand on generating capacity. This problem could become a significant challenge in some parts of the country unless new facilities are constructed or demand is reduced. In the meantime, electricity will retain its 10 rating for consumer convenience.

Appliances and Equipment

Gas enjoys the widest selection of high-efficiency furnaces and boilers of any energy source. It also ranks high in choices of hot water appliances, cooking ranges, clothes dryers, and industrial equipment. However, it lacks adequate selections of cooling and refrigeration equipment. Major manufacturers offer the most advanced equipment for gas-fired cogeneration and combined cycle electricity generation. Factory-built natural gas vehicles are not yet available, but they may be marketed in the near future. Overall, a positive rating of 7 seems appropriate.

For stationary applications, oil has a fair supply of boilers and furnaces, though choices are more limited than for gas. Industrial equipment often has dual-fuel capacity, enabling the user to switch from gas to oil or vice versa. In the transportation sector, oil has a virtual monopoly in vehicles using gasoline or diesel fuel and in filling stations servicing them. If gas (or any other fuel) wants to get into the transportation market in a big way, it would be advisable to join forces with the oil products distribution network. In the stationary area, the limited availability of oil burning appliances would give oil a low rating of 3, but in the transporation sector it deserves a rating of 10. Taking all factors into consideration, we will assign it 7.

To keep operating expenses under control, coal uses a great deal of sophisticated equipment. The challenge is intensified by the need to minimize pollution. Because the livelihood of many people is involved with coal production and utilization, this fuel enjoys strong political support. Government funds are available for the development of more efficient and less polluting coal technologies. A rating of 7 for equipment seems appropriate.

Electricity enjoys the strongest support of appliance and equipment manufacturers. In most fields, the product lines offered for use with electricity dwarf those for competing energy sources, including gas. Electricity dominates lighting, air conditioning, television, computers, refrigeration, and many other fields. However, gas appliances can more than hold their own for space heating, hot water production, cooking, clothes drying, and large-scale air conditioning. The strong support by appliance manufacturers gives electricity a rating of 10.

Versatility

Natural gas is the most versatile of all energy sources. It can be used for space heating, hot water production, cooling, cooking, drying, cogeneration, and combined cycle electricity generation. It has numerous industrial applications, including many high technology procedures. In compressed form, it is an excellent vehicular fuel. In can also play a role in improving the combustion of other fuels, such as coal. It performs all of these functions in an environmentally benign fashion. The 10 rating is well deserved.

Oil's versatility is similar to that of gas. It can be used in many of the applications cited for gas, but it is inappropriate for some direct industrial and commercial uses (e.g., food processing). Oil products enjoy a virtual monopoly of transportation markets. The versatility of oil gets an 8 rating.

Coal has limited versatility. It is used primarily for generating electricity or industrial power. In the past, it also played a role in rail transportation, powering steam locomotives. However, that function is no longer significant. From a practical standpoint, coals versatility rating is a low 3.

Electricity has a wide range of stationary applications, but its use as a vehicular energy source is limited. The development of highly efficient batteries and other new technologies, such as superconducting devices near normal temperatures, would improve its role in mobile markets. Until such breakthroughs occur, its versatility rating is 7.

Energy Security

Over 90 percent of the natural gas used in the U.S. comes from domestic sources; the rest is imported from Canada. The long-term outlook for gas supplies from domestic sources remains favorable. Gas has the advantage of being readily substitutable for oil in many applications. In terms of energy security, gas qualifies for a 10 rating.

Oil imports account for more than forty percent of U.S.

requirements. In coming years, imports are likely to grow even more, while domestic production declines. Oil is a weak link in terms of U.S. energy security. A low rating of 3 calls attention to U.S. vulnerability in this category.

The U.S. has enormous coal resources, which are adequate to take care of foreseeable requirements for centuries. In fact, the U.S. exports considerable amounts of coal. However, coal production is vulnerable to disruption by strikes. Moreover, coal deliveries may be adversely affected by weather conditions. The energy security rating of coal is 9.

Electricity uses primarily coal, nuclear energy, and hydroelectric power, all of which are in adequate supply from domestic sources. In some areas of the country, notably the Northeast, oil plays a role in electricity generation, which keeps the security rating at 9.

Public Image

Natural gas has the best public image of any energy source. It is identified with environmental cleanliness, which gives it a big advantage in public ratings. The fact that it is available from plentiful domestic sources also supports a positive image. Most gas utilities have done a fine job keeping their customers satisfied with good service at reasonable prices. Government officials, regulators, and energy decision makers have increasingly recognized the many advantages of natural gas, which can play an important role in helping to solve energy and environmental problems. A rating of 10 is well deserved.

Oil's public image dropped precipitously as a result of the price escalations of the 1970's. This experience has led to considerable distrust of the oil industry and of oil as a fuel. Moreover, oil is on the defensive environmentally, particularly in the transportation sector. Gasoline and diesel fuel are held primarily responsible for ozone (smog) pollution in many urban areas. Oil tanker mishaps have added to the low public image, which justifies a rating no higher than 3. This low rating is particularly significant in view of the fact that throughout most of the twentieth century, oil was the unchallenged energy leader.

Coal's public image is low primarily because of its role in causing acid rain and other types of pollution. The development of clean-burning technologies, including the select use of gas with coal, should facilitate a more positive image for coal in the future. In the meantime, it has a low rating. of 4.

Electricity has been affected by the environmental problems caused by coal as well as by the public's antipathy to nuclear facilities. Most electric utilities have done a good job maintaining a fairly satisfactory public image in spite of these handicaps. A positive rating of 7 reflects good public relations management.

The following table summarizes the ratings data.

Ratings of Energy Sources

Criteria	Gas	Oil	Coal	Electricity
Production costs	8	5	5	2
Transportation costs	6	8	6	8
Environmental costs	10	4	1	3
Consumer values	10	5	7	7
End use convenience	10	7	1	10
Appliances and equipment	7	7	7	10
Versatility	10	8	3	7
Energy security	10	3	9	9
Public image	10	3	4	7
Total	81	50	43	63

When all factors are added up, gas ranks first, with 81 points out of 90 possible. Natural gas is 18 points higher than electricity, 31 points abouve oil and 38 points ahead of coal. It is the only energy source with six perfect scores (environmental costs, consumer values, end use convenience, versatility, energy security, and public image). These high ratings support the conclusion that natural gas is the best energy choice.

Sources:

[1] 1988 Gas Facts, *American Gas Association.*

[2] *"Debate Begins on Bush's Clean-Air Plan," by Rose Gutfeld,* Wall Street Journal, *July 24, 1989.*

About the Author

Ernest J. Oppenheimer has done extensive research and writing about energy. His studies led him to the conclusion that natural gas could play a significant role in solving the energy problem. His previous book, *Natural Gas: The New Energy Leader,* published in 1981, presented this thesis. Since that time, many developments have taken place which have enhanced the importance of natural gas. The environmental benefits of this clean-burning fuel have become increasingly recognized. New technologies for gas cooling, cogeneration, and combined cycle electricity production, have emerged. The domestic petroleum industry has shifted its primary focus from oil to natural gas. Internationally, natural gas has made major strides forward. All of these factors, plus many others, have contributed to the decision to write a completely new book about natural gas. The present volume is the result of this effort.

The author has also written three other books on energy-related topics, as well as a book on inflation. Prior to his writing career, Dr. Oppenheimer spent fifteen years doing research and consulting work in the investment banking field. Much of his research concerned technological industries and companies.

He received the doctor of philosophy degree in the social sciences from the University of Chicago. His studies included international relations, economics, political science, and social psychology.

Dr. Oppenheimer has been interviewed on television and radio talk shows and has lectured to business groups and to university audiences.

Books by Ernest J. Oppenheimer, Ph.D.

Natural Gas, the Best Energy Choice
Gasoline Tax Advantages
Solving the U.S. Energy Problems
Natural Gas: The New Energy Leader
A Realistic Approach to U.S. Energy Independence
The Inflation Swindle

Some Gas Statistics (1987)*

Number of meters:	
Residential	47,362,400
Commercial	3,979,600
Industrial	180,000
Other	53,600
Total	51,575,600

It is estimated that 172 million Americans are served by natural gas utilities.

Gas appliance shipments by manufacturers	
Central heating	2,263,000
Ranges	2,143,000
Water heaters	3,673,000
Clothes dryers	1,037,000
Revenues of gas utilities	$45,491,924,000
Gas utility plant	$93,568,000,000
U.S. energy consumption (trillion Btu):	
Oil	32,627
Coal and coke	18,029
Natural gas	17,180
Nuclear power	4,920
Hydro power	3,041
Other	245
Total	76,042

Source: 1988 Gas Facts *(1987 data), American Gas Association, 1988.*

Information Sources

American Gas Association, 1515 Wilson Boulevard, Arlington, Virginia, 22209. Telephone: 702-841-8400.

Gas Research Institute, 8600 Bryn Mawr Avenue, Chicago, Illinois 60631. Telephone: 312-399-8100.

Natural Gas Vehicle Coalition, Two Lafayette Centre, 1133 21st Street, N.W., Washington, D.C. 20036. Telephone: 202-466-9038

Energy Measurements of Natural Gas

1 Cubic Foot (cf) = 1,026.9 British thermal units (Btu).
1 Mcf = 1,000 cubic feet = 1,026,900 Btu.
1 MMcf = 1,000,000 cubic feet = 1,026,900,000 Btu.
1 Bcf = 1,000,000,000 cubic feet (one billion cubic feet).
1 Tcf = 1,000,000,000,000 cubic feet (one trillion cubic feet).
1 Tcf = 1,026,900,000,000,000 Btu (one quadrillion Btu, or one Quad).

One British thermal unit (Btu): the amount of heat required to raise the temperature of one pound of water by one degree Fahrenheit.

1 Therm = 100,000 Btu.

Metric Conversions
1 Cubic Foot = 0.02832 cubic meters
1 Therm = 25,200 Kilocalories
1 Mcf = 28.32 cubic meters

**Source:* 1988 Gas Facts, *Appendix B, American Gas Association, 1988.*

Index

Order Form

Natural Gas, the Best Energy Choice, (hard cover edition) by Ernest J. Oppenheimer, Ph.D.

__________ copies at $22.50 each $ ____________

PAY FOR THREE, GET ONE FREE
(applies only to the above book)

The following previously published paperback books by Dr. Oppenheimer are also available.

A Realistic Approach to U.S. Energy Independence (1980)

__________ copies at $5.00 each $ ____________

Natural Gas, the New Energy Leader (1981)

__________ copies at $7.50 each $ ____________

Solving the U.S. Energy Problem (1984)

__________ copies at $5.00 each $ ____________

Gasoline Tax Advantages (1987)

__________ copies at $10.00 each $ ____________

SPECIAL COMBINATION OFFER: all five of the above books, a $50 value, for $40.

__________ sets of five each for $40.00 $ ____________

Subtotal $ ____________

Residents of New York State add sales tax $ ____________

Shipping expenses (see next page) $ ____________

Total $ ____________

(Continued on next page)

Shipping expenses: No additional charge for U.S. destinations. For Canada, add ten percent of the subtotal. For destinations in Western Europe and Latin America, add twenty percent. For the far East and Eastern Europe, add thirty percent.

Please enclose payment with order. Make check payable to Pen & Podium, Inc. Checks from foreign countries should be in U.S. currency, drawn on a bank situated in New York City. Send order to:

Pen & Podium, Inc.
40 Central Park South
New York, N.Y. 10019
Telephone: (212) 759-8454

Please print or type your name and address below:

Name ______________________________

Company ______________________________

Street ______________________________

City, State, Zip ______________________________